Rosemary [illegible] 1977

pp. 45, 46, 53 are important

YOUNG HORSE MANAGEMENT

Other books by the same author

THE AMATEUR HORSE BREEDER

STALLION MANAGEMENT

YOUNG HORSE MANAGEMENT

A. C. Leighton Hardman

PELHAM BOOKS

First published in Great Britain by Pelham Books Ltd
52 Bedford Square, London WC1B 3EF
1976

ISBN 0 7207 0904 0

Set and printed in Great Britain by
Tonbridge Printers Ltd, Peach Hall Works, Tonbridge, Kent
in Baskerville eleven on thirteen point on paper supplied by
P. F. Bingham Ltd, and bound by Redwood Burn
at Esher, Surrey

CONTENTS

ILLUSTRATIONS

DRAWINGS

ACKNOWLEDGEMENTS

I should like to express my gratitude to the following people: Sidney Ricketts M.R.C.V.S. for the very great help he has given me in reading, correcting and up-dating large sections of this book; David M. FitzGerald for his helpful advice relating to the legal aspect of selling horses and also for thinking of a suitable title; John Hancock of Bloodstock and General Insurance for reading and correcting Chapter 16; Peter Bolton for his advice on the latest agricultural trends; Johanna Vardon of the National Foaling Bank for advising me about the rearing of orphan foals; and especially Peter Faulkes for the use of the beautiful photograph on the jacket cover.

NEWMARKET
JANUARY 1976
A.C.L.H.

FOREWORD

By Alan Lillingston

In asking me to write a foreword to her book, the author has paid me more of a compliment than perhaps she realises.

This book, I believe, should have a great future, and occupy a leading position amongst reference books on the subject of horse breeding.

It will appeal not only to the novice, as it deals most clearly and concisely with the elementary principles of breeding, but also to the expert as it contains a fund of information based on practical experience and intensive study.

In general, success in bloodstock breeding depends upon careful selection of the individuals involved, and then great attention to detail in the rearing of the offspring. The author's experience in the showring and as a judge, give her an excellent background to select for conformation. The manner in which her animals have been prepared for Tattersalls sales is proof indeed to me that she is a master of the art of rearing and preparing thoroughbreds for the sale ring.

At this time of general economic crisis, which is aggravated further in the bloodstock industry by enormous overproduction, only the most efficient commercial breeders with a truly professional approach are likely to survive. This book provides, in the greatest detail, many of the 'tricks of the trade'. If studied carefully and referred to when necessary it should be of tremendous value to all breeders.

I

NOTES ON ESTABLISHING AND RUNNING THE STUD

Many people who breed horses on a few acres dream of owning their own stud. Few ever realise this ambition, but for those who do several factors have to be taken into consideration. The one which will most influence the success or otherwise of a horse-breeding or rearing enterprise is probably its location. As the type of soil, climate and region cannot be altered, careful thought must be given to these points before establishing a new stud.

For ease of marketing, studs must be located near to or in a horse-breeding area. Failing this they must be near a motorway or major trunk road. Private buyers are notoriously reluctant to travel any distance to buy horses when they can find what they want within a small radius of a horse-breeding centre.

A light chalk or sandy soil is desirable from the point of view of its dryness and consequent resistance to poaching. Also this type of soil is seldom very fertile, and therefore gives a comparatively low response to any fertiliser treatment, with consequently less chance of leg troubles due to nutritional factors. This type of leg trouble (round joints, etc.) can be precipitated by a sudden change in a foal's or yearling's diet to a higher protein level. It can occur when these horses are turned on to pasture in the spring which has been boosted with artificial nitrogen fertiliser. Organic fertilisers are released at a slower rate than artificials, and are therefore less likely to produce a rapid growth of grass and so upset the bowel flora and general metabolism of the horse, but are very much more expensive.

In these days of soaring inflation, it is becoming increasingly necessary for studs to rely on home-produced foodstuffs. Grass is still the cheapest. Therefore, although management problems may be greater, it is desirable, if only from an economic point of view, to establish a stud on good fertile land, preferably in a mild, limestone area, with a comparatively high rainfall. Low rainfall areas are usually subject to a few weeks drought during the summer each year, when the grass burns up and hay must be fed. When this happens the ground tends to become as hard as concrete, which is not good for the legs of young horses and necessitates bringing them into a straw yard or loose-boxes, or alternatively discing and harrowing a paddock down to a fine seed-bed, and using this as an exercise area until the ground becomes softer. In high rainfall areas this seldom occurs and the grass remains green throughout the whole year.

Paddocks which are to be used for horses, particularly youngstock, must be as flat and level as possible to avoid injuries. Land which slopes towards a fence is very dangerous, as horses can get injured if they are galloping downhill and the ground is too slippery for them to be able to pull up at the bottom.

The management system to be used will to a large extent influence the size and location of the paddocks. Small paddocks – 6 acres or less – are more economical to manage than larger areas due to the selective grazing habits of horses and the need to catch and handle them individually. Smaller paddocks give room for the animals to be moved around the stud without the necessity of leaving them in any one paddock for more than a month at a time and thus avoiding the risk of the land becoming 'horse sick'. Under this system cattle can follow the horses and so help to keep the land clean and evenly grazed. Mixing the species of animal will also help to reduce the incidence of disease and unthriftiness usually found where there is a high continuing concentration of a single species in a given area. This applies equally to the land and buildings. Only relatively mature cattle should be used for grazing with young horses; calves and young steers compete for the available minerals in the soil, which are depleted at the expense of the horses.

Where possible existing trees and hedges should be retained when laying out the paddocks, as shelter belts are very important and take some considerable time to become established. Both grassland and the grazing animals do better where there is plenty of shelter – open windswept paddocks will not produce well-grown youngstock or an early bite. However where no shelter exists, quick-growing non-poisonous varieties of trees and shrubs should be planted, at sufficient distance outside the paddock fence so that even the largest horses cannot reach them. Copper Beech makes an ideal hedge, as it gives shelter both summer and winter. It looks beautiful and is non-poisonous, but rather expensive. Ideally, shelter belts should be established on all four sides of the paddocks, but failing this at least against north and east winds. Close-boarded fencing should be erected in particularly exposed places until the trees have become established.

Due to the fact that horses tend to poach wet land badly, paddocks must be well drained. From the point of view of safety it is better to tile drain all open ditches and fill them in completely. This will prevent accidents, particularly where young foals are concerned. Failing this, ditches and ponds should be very well fenced with at least 3 and preferably 4 or more rails, as foals will often climb through or roll under a 2-rail fence. Post and rail fencing or stock-proof hedges are the only really safe forms of fencing for young horses. If wire has to be used for any reason, it must be kept absolutely tight – loose or slack wire is highly dangerous. Double fencing will help to prevent horses leaning on the fence in an effort to reach their neighbours on the other side and the corners of paddocks should be rounded to prevent any animals from being cornered.

To save labour, mains water should be laid on to all the paddocks. The troughs should be fitted with automatic filling devices and have a plug in the bottom to facilitate regular cleaning. Water troughs should be sited well away from gateways to avoid a sea of mud in wet weather. As far as possible they should not be placed directly under trees, otherwise they will fill up with leaves, especially during the autumn. But at the same time they should be installed in a sheltered position.

To get the best return from grassland, the soil must be in good condition as well as being well drained. This means there must not be any deficiencies of lime, phosphates, potash or nitrogen. Excess lime can lead to deficiencies of other minerals such as phosphorus, with resulting bone troubles. Lime, phosphates, potash and nitrogen are all washed out of the top soil at varying rates over a period of time. Therefore regular soil analyses should be carried out – about once every 3–5 years – and any deficiency which becomes apparent should be corrected. Normally, many of the nutrients are replaced by the grazing animals, but where the droppings are picked up from the paddocks regularly, deficiencies may occur. A phosphate deficiency, for example, can cause soft bones which in turn may result in splints, etc. Phosphates are essential for good root development in the plants and therefore benefit the growth of clover, which in turn helps the nitrogen level. Potash is necessary for healthy plants.

Nitrogen is leached out of the soil at a far greater rate than any of the elements already mentioned, and needs renewing annually. High applications of nitrogen will lead to heavy crops of grass, with high protein levels, quite unsuitable to the needs of grazing horses although necessary for good hay crops. Where an early bite is required on studs, it is possible to boost the grass but keep the extra growth down to a minimum by only applying a light dressing of nitrogen fertiliser, not exceeding 45–50 units of nitrogen. Where clover is well established in the sward it will supply about 150 units of nitrogen per season. To get the best response to fertiliser treatment it should be applied in the early spring once the grass has begun to grow, when it can be utilised to maximum advantage. If applied at a time of slow growth, a percentage of the fertiliser will be washed out of the top soil before it can be used by the plants.

Ideally, an application of 10 tons to the acre of ordinary farmyard manure each autumn – not poultry manure, which is too high in nitrogen – will supply all the nutrients, apart from lime, necessary to establish and maintain stud paddocks in optimum condition. In the early spring the paddocks should be harrowed, ideally using Parmiter or ripper type harrows, to pull out the dead grass and spread any remaining manure. After harrowing, the paddocks should be rolled with heavy flat grass rollers – the Cambridge variety are too light for most grassland. Throughout the spring, summer and autumn, when the land is not too wet, the paddocks should be topped, chain harrowed and rolled each time between grazings. Harrowing and rolling, in particular, should be carried out slowly for maximum efficiency. Topping is essential when mixed grazing is not practised, to maintain an even growth rate of the grass and the maximum nutritional value and palatability.

When selecting grass mixtures for paddocks, the following points should be taken into consideration:

Horses are very selective grazers. Therefore palatability is of great importance and strains of grass should be selected for paddocks with this in mind. Nearly all grasses are palatable when young, but some become less palatable with age. Leafy varieties have a higher protein content, with lower fibre, and are therefore more digestible. Cocksfoot, except when very young, is too coarse and tends to be rejected by horses. Soil types and local climatic conditions will influence the actual varieties and quantities of each used in the mixture; and your local Agricultural Development Advisory Service (A.D.A.S.) officer (in England) should be consulted when in doubt.

Horses have the habit of cropping grass almost to ground level. Care should therefore be taken not to include too much clover in any paddock mixture, as under heavy grazing conditions this will tend to grow away at the expense of the grass – providing a lower overall yield of herbage per acre. A mixture containing several different varieties of grass, as well as herbs and clover, is ideal for horses.

Paddock grasses must be able to withstand poaching to a large degree. Therefore they must have good tillering ability – that is, the ability to give off new shoots at their base and thus spread out at ground level. Grasses which exhibit speed of recovery after grazing are also very desirable.

If at all possible, horses should not be turned out on to newly-established paddocks in their first year, unless the ground is hard. Some people graze them with sheep first, but this presents the problem that sheep netting has to be erected all round the paddock. A cut of hay can be taken, but this is not recommended as an annual practice, as it tends to spoil the tillering ability of the grass. But it is desirable, these days, to set aside a certain acreage for hay production and sow these paddocks with recommended varieties of rye-grass and clover. This is a 1-year ley and therefore requires re-sowing every year. Lucerne presents an excellent alternative where adequate drying facilities or climatic and soil conditions exist and will last for 5 years, but due to its high protein content should be fed to mature horses mixed with meadow hay. These are the best mixtures for horse hay.

You will need to allow about 1 acre per horse, to provide one cut yielding on average from 150 bales to the acre in the N.W. to 90 bales to the acre in the S.E., and allowing 15 lb.

of hay per horse per day, i.e. about 1/3 of a 40 lb. bale. With luck you should be able to get 2 cuts of hay per year, the second cut yielding 2/3 the weight of the first cut. This would then be available for sale or for cattle feed. Similarly, if the acreage permits, it would be a good idea to grow your own oats, yielding an average of 35 cwt per acre, which in turn would satisfy the winter requirements of about 1 horse and provide 1 ton of straw, which could be traded for good wheat straw or fed to cattle.

Wooden gates are probably safer than metal gates, as they will tend to break if a horse puts a leg through the bars. Where metal gates are used, they should have their bars running vertically and sufficiently far apart so that a horse can pull its leg clear in the event of an accident. Wire mesh (Weldmesh) metal grilles can be fitted to both wooden and metal gates to make them safer, and wooden gates last longer if they have metal bands on all exposed edges, to prevent horses from chewing them but care should be taken to put the metal on in short strips so that it will 'break up' in the event of an accident. Ease of opening and closing gates is essential on a stud, and a wicket-type gate fastening is ideal. To make this horse-proof, a chain should also be fitted which fastens with a large staple and metal pin, the latter also on a short chain so that it will not get lost.

Gates are best located with direct access on to a hard road and the immediate area inside the paddock must be very well drained to cut poaching down to a minimum. They should not be constructed in the corner of a paddock, as horses tend to crowd round gateways when it is time to come in, and plenty of room for escape must be allowed in case kicking matches develop.

When laying out the stud and constructing buildings the following points must be taken into consideration:

(1) The length of time a particular building is likely to last. With inflation at its current annual rate of growth, it really does pay to build now with an eye to the future. To

this end, concrete block or brick construction is better than wood. American barn-type buildings with single span roofs housing 4 rows of boxes reduce the overall cost of building construction considerably, although adequate ventilation and fire precautions must be fully recognised in the design.

(2) The cost of maintenance, sub-divided into labour and materials. Low-maintenance buildings are essential. High-maintenance buildings can run away with 25 per cent or more of the total running costs of the stud.

(3) The overall daily labour requirement of the units. This should be kept to a minimum. In the future, to cut the cost of production, mechanisation must become an important factor and time must be spent at management level working out how the labour requirement can be kept to a minimum and better use made of available equipment. To this end, buildings must be constructed with sufficient width between them to allow for the easy passage of tractors and their equipment, not forgetting extra width at corners, to allow for turning. For example, it is possible to reduce the time and work force involved in mucking out by constructing a central pit, and by pushing the day's muck thrown out of the boxes down to the pit with a tractor-mounted scraper. Manure pits should be constructed with double-buttressed walls, so that they will not collapse if hit by a tractor. Thought should also be given to forms of deep litter bedding which would eliminate daily mucking out and the massive labour requirement for this operation.

As far as possible the various classes of horses – broodmares, weaned foals and yearlings – should be housed in separate units, which will help to avoid the spread of disease and facilitate management. Where individual units are large enough, it is a good idea to put one person in charge of each and make them responsible for that unit. This method of management engenders pride of work among the staff, which is often lacking where hands work generally throughout the whole stud. The person in charge should be consulted when any change or addition to that unit is being considered.

There are two ways of designing a stud:

(1) Create individual units of boxes combined with parcels of land suitable for the economic running of those units. These can be located some distance from each other, and should allow at least one acre per animal.

(2) Build all the units close together, with central feed store, Dutch barn, exercise yard and manure heap.

From the point of view of economy in building, the second method is probably the best; but from the point of view of turning out, the first method is less time-consuming. With either method, an isolation unit should be built some distance from the other buildings. On small studs this need only be one or two boxes, with adjoining paddocks, but on all studs of any size it is very necessary to have somewhere to isolate sick horses and those which have come from abroad or been in contact with possible disease.

Loose-boxes which are designed specifically for young horses must be so made that they discourage the development of stable vices. They should be as warm as possible without reducing the amount of ventilation. Warmth is a particularly important factor when foals and yearlings are being prepared for sale. In the British Isles the major yearling sales are held in September and October, so the yearlings' full summer coats have to be maintained until then. The loose-boxes should therefore be sited in a sheltered position with their backs to the north-east and must be fully lined and insulated.

As far as stable vices are concerned, the most important single vice, which must be discouraged at all costs, is crib-biting. In order to practise this vice the horse must be able to catch hold of some projecting object with his teeth, therefore smooth cement-rendered walls are the ideal deterrent. Young horses should be able to see out of their boxes, and top doors should be kept open as much as possible for maximum ventilation and to prevent boredom. As far as possible an interesting but not too exciting view should be provided for their entertainment. A completely blank wall opposite the loose-box doors does tend to create boredom.

2

MANAGEMENT OF THE YOUNG FOAL

With the ever-increasing cost of keep and foaling charges on public studs, the general trend is for more and more people to foal their mares at home and only take them to the stallion when they are ready for covering. In areas where experienced veterinary help is readily available and adequate facilities exist on the mare owner's stud, there is nothing to be lost by foaling mares at home. In all other circumstances mare owners should consider the pros and cons very carefully. Unless adequate facilities exist the mare owner is merely gambling the value of the unborn foal and his mare against the added cost he would incur by allowing the mare to foal on a public stud where there is every possible facility.

When a mare is due to foal, it is essential that someone sits up all night, both before foaling and preferably for the night after, to make certain that the mare and her foal are under supervision should any complication arise, such as a difficult foaling, haemorrhage in the mare or meconium colic in the foal.

As soon as the foal is born it must start to adjust to its new way of life and environment. With the normal healthy foal this does not usually present any difficulty, but in the case of a sick foal, skilled veterinary help might well be essential for its survival.

The newborn foal must have started to breathe by the time the cord is severed. If the cord is broken too early it will deprive the foal of a considerable quantity of blood and lead to anaemia or may even help to produce a 'Barker' foal (see page 37). Normally the cord will break about $1\frac{1}{2}$ in. from the umbilicus as the foal struggles to get up, or when the mare gets to her feet.

It can be broken by the attendant *once the pulsations have stopped,* by dragging the foal round to the mare's head. This is a better method than cutting the cord, since it is less likely to lead to any infection, as the stretching and re-coil of the vessels produce a more natural seal. If bleeding from the stump does occur it is usually stopped, quite easily, by pinching the end between the fingers for a few minutes. The end of the stump must be dusted liberally with an antibiotic powder to help prevent infection (see Navel Ill, page 44).

On many studs heat lamps are used to raise the temperature of the foaling box, although some authorities maintain that the sooner a foal hardens off the better, and never use artificial heating except for sick foals or in very cold weather. However, heat lamps must be available in all foaling boxes for immediate use should the necessity arise. It is better to foal mares inside from the point of view of keeping an eye on them, particularly if the weather is cold and wet, as a sudden dramatic change in temperature from the womb, and lack of grease in the foal's coat at this time, can lead to pneumonia.

If it is decided to foal a mare outside, great care must be taken to ensure that the paddock is absolutely safe, with the rails low enough to prevent a newborn foal from getting under the fence. There should be no streams, open ditches or ponds into which the foal can fall. Newborn foals will scramble some considerable distance across a paddock while struggling to get to their feet, and during this period can quite easily slip under a fence or fall into a ditch. When mares foal outside unattended, there is also the risk that one may get into difficulties or even savage its own foal; or another mare may do so before the foaling mare has time to get to her feet to protect her offspring. In the British Isles adverse weather conditions during spring and early summer can also be a problem.

As soon as delivery is complete,* when the cord has severed and the mare can reach her foal, she will usually lick it all over. By doing this she is learning to recognise her own foal from the taste and smell of the fluids in which it is soaked. Where the birth is attended, assistance can be given to dry the foal off, by rubbing it down with a *clean*, dry, chamois leather which has been rung out in warm water first, or with a dry *clean* towel. In each case it is important that the cloth should not smell strongly of other horses, which would only tend to confuse the mare.

Before birth, the foal is in relatively sterile conditions, but at birth it suddenly comes into contact with every germ in its immediate surroundings. In order to survive it must develop immunity to these organisms as quickly as possible. Unlike the human baby, the foal does not receive any resistance to disease before birth, but has to acquire all its immunity from the colostrum or fore-milk in the first 24 hours after birth. The level of antibodies in the colostrum can be specially raised against tetanus and influenza. The former can lead to the death of horses of any age and the latter can be a killer where young foals are concerned. To do this the mare is vaccinated one month before she is due to foal. If the mare is not vaccinated against tetanus, then tetanus antiserum should be given to the foal soon after birth and again a month later.

Young foals travelling to public studs, possibly in public transport, and maybe as young as six days old, are highly 'at risk' to infection. It may therefore be considered good

*Detailed chapters on foaling mares can be found in *The Amateur Horse Breeder* and *Stallion Management* A. C. Leighton Hardman, Pelham Books.

policy to protect them by antibiotic injection for the first three days of their lives. By the time a foal shows obvious signs of illness, serious damage can have been done, as they have little resistance to disease as compared with older horses. Prevention is better than cure.

When dealing with valuable foals and where time permits, it is a good idea to take the foals' temperatures as a routine practice before turning out each morning. In this way any slight rise in temperature can be noted and treatment given before serious infection becomes established.

The colostrum or fore-milk contains high levels of antibodies, substances found in the blood which destroy germs that would otherwise cause disease. But a foal's ability to absorb these across the small intestine ceases after about 24 hours. After this the foal can only obtain immunity artificially, by means of injections, or by natural development over a period of time. Where a mare has been running her milk before foaling, the antibody level will probably be low, so an alternative supply of mare's colostrum should be found.

It is a wise precaution to take a little colostrum, after foaling, from every mare which has not run her milk (as long as she has not produced a haemolytic foal). If you only possess one mare yourself collaborate with mare-owning friends to build up a colostrum bank between you, as you never know when you may need some urgently. Colostrum kept in a deep-freeze will keep from the end of one season until the beginning of the next and so be available for emergencies. The colostrum should be put into sterile polythene bottles and dated with a waterproof felt marker pen, before being placed in a deep-freeze. To sterilise the bottle use boiling water, a sterilising agent recommended for baby's bottles, or dairy hypochlorite.

Ideally, the foal should stand and suck within two hours of birth. In practice, the length of time varies enormously between individuals. Most foals show a marked suck reflex within five minutes of birth. They will tend to suck the walls

If the foal has not done a dropping by the time he is 12 hours old, give a tablespoon of liquid paraffin (in a warming syringe)

of the loose-box or the mare's legs even before they are able to get up, and for this reason mud should be removed from the legs of foaling mares.

Foals will struggle to get up and fall many times before they finally manage to get to their feet. Periods of activity are usually divided by periods of rest before the foal finally gets control of its legs and stands. Unless the foal is weak or the mare is being awkward, it is much better to leave the foal to get up on its own, making sure that there is sufficient straw on the floor of the loose-box to stop the foal injuring itself.

Once up, the foal will be drawn by instinct to the darker areas under the mare's belly and between her legs. The mare can be encouraged to let her milk down by swabbing her udder with a piece of cotton wool soaked in hand-hot water; this is particularly helpful where maiden mares are concerned. A little Vaseline can also be smeared on the teats of maiden mares. As long as the mare co-operates the foal will gradually work towards the udder, gaining strength and balance by the minute, until it finds a teat; this can take upwards of an hour.

Unless the foal is weak or the mare ticklish, it is unnecessary to offer any assistance to the foal. Physically forcing the foal's head under the mare and his mouth towards the teat will only confuse him and may delay the actual start of sucking. If the mare insists on following her foal round the loose-box, it is better to hold her head. If she attempts to kick the foal, it may be necessary to hold up one of her front legs or take a handful of loose skin on her neck to restrain her, but *never put a twitch on a newly-foaled mare.* This can cause the blood pressure to rise and could lead to haemorrhage into the ligaments of the uterus so soon after foaling.

The earlier a foal is taught to lead the easier it will be. A first-size foal headcollar can be put on the foal the day after it is born; this must be clean and soft to prevent it from rubbing the tender skin of the baby foal. It need not be very expensive, as this size of headcollar is usually only worn for about 2–3 months, but a word of warning here: all headcollars should be made from a material which will break under stress – far better a broken headcollar than a broken neck. If the weather is fine and warm, there is absolutely no reason why the mare and foal should not be turned out into a small sheltered paddock for a few hours the next day. It can be very dangerous to let young foals run loose round the stud behind their mothers; therefore some means of restraint should be used until the foal will lead quietly in-hand. The best method of leading very young foals is for two people to hold a sack round the foal's hind-quarters and link hands across its chest (see Fig. 1). In this way a young foal can be guided anywhere in complete safety, the mare being led by another person behind the foal.

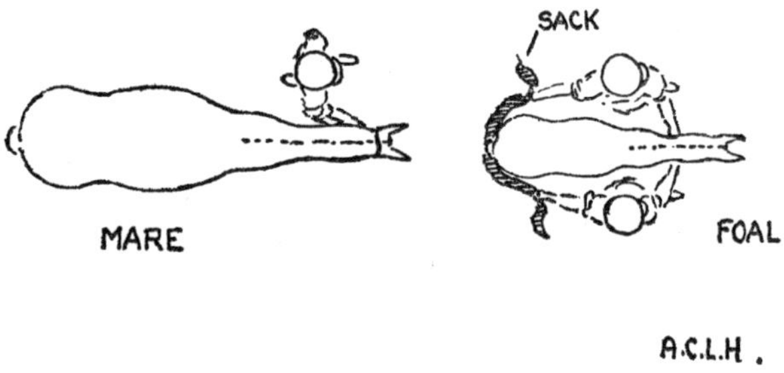

Fig. 1. How to guide a young foal without the use of a headcollar

The foal should be taught to lead from its headcollar as soon as possible; this is best done in the loose-box. When putting the headcollar on, make sure that it fits reasonably tightly so that the foal cannot get a hoof through a strap, but not so tight that it will rub. As foals grow very rapidly and soon outgrow their headcollars, someone on the stud should be made responsible for checking at least once a week, to make sure that all the headcollars still fit correctly.

For the initial lesson in leading, slip a rein through the

headcollar, get someone to lead the mare slowly round the loose-box and follow with the foal. The first time the foal feels a pull on the headcollar it will almost certainly run backwards, and may rear or fling itself on to the ground, so a soft landing is essential. For this reason, never attempt to lead a foal for the first time in a concrete yard.

In order to encourage the foal to move forward freely, place your right hand round the hind-quarters; do not twist the tail round. Keep your left hand on the rein and guide the foal round the loose-box behind the mare (see Fig. 2). Once it gets the correct idea, remove your right hand and hold the base of the foal's neck. You will be surprised how much control this gives you and what a steadying effect it has on the foal. Young foals should never be knocked about or treated roughly, otherwise they may turn sour and become temperamental.

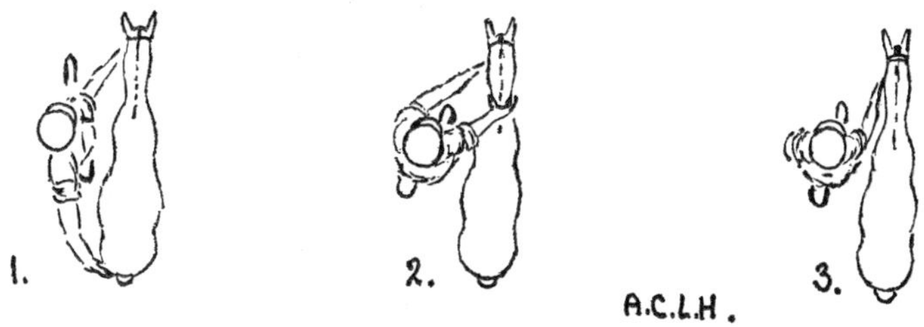

Fig. 2. The three stages in teaching a foal to lead from a headcollar

If a foal is lying down and you wish to get him up on to his feet, the easiest way of doing this is to straighten his front legs out and then run your hand down his back bone. In the majority of cases, you will find that the foal will immediately leap to his feet without any further trouble. Should you ever have occasion to lift the foal to his feet, never grab him with both hands under the stomach, as you may injure him. Rather put one hand round his chest and the other round his hind-quarters and lift. It sometimes happens that a newborn foal is frightened to lie down and will remain standing even when it is obviously sleepy. In this case get hold of the foal as described on page 31, lift it off its feet with its legs towards you, and lay it down gently on the straw.

A large covered yard is a great asset to any stud and might even be justified on a small private stud. The possession of such a yard means that all stock, including mares and young foals, can be given some exercise every day, regardless of the weather, and eliminates the necessity of poaching any paddocks during wet winter or early spring weather.

If the weather is suitable, mares with young foals can be turned out in a small, well-fenced, sheltered paddock from the day after birth. The mare should be walked round for about 10 minutes before she is finally loosed; in this way the foal will get used to following its mother. Allow the mare to put her head down and graze and as soon as she seems settled, slip the lead rein. If on the other hand the weather is cold and wet for a few days and you cannot turn them out, it is essential – unless you know she is very quiet – to gallop the mare on her own in a small paddock *before* you let her loose with her foal. Mares which have been confined to a loose-box, for any length of time, will tend to gallop off, kicking their hind legs in the air the first time they are turned out. This can be very dangerous; young foals have been killed or had their jaws fractured by being too close to the mare when she kicked out.

To gallop a mare, use a very well-fenced small paddock – a close-boarded stallion paddock is ideal. Failing this, get a sufficient number of people to stand round the outside of the fence, so that they can turn the mare as she gallops towards them. As soon as the mare is tired and stops galloping she can be led back to her foal, which has been left – preferably held by somebody – in its own loose-box with the top door closed. Then both the mare and the foal can be led out. The mare should be walked round the paddock and allowed to graze until she appears settled; then the lead rein can be slipped from the headcollar.

Where there are several mares on the stud foaling at or

around the same time, it is advisable to turn the mares and foals into their paddock singly, allowing each mare and foal to settle before the next is turned out. Otherwise you will run the risk of getting the foals mixed up, since young foals have a tendency to gallop off on their own and will often run to other mares when they get lost. For this reason it is better not to turn too many mares into a paddock together until the foals are nearly a month old.

For ease of operation a standard method of catching foals should be adopted and adhered to at all times. Any method which works can be used, but well-tried and satisfactory methods are as shown in Figs. 3 and 4. Eventually you will

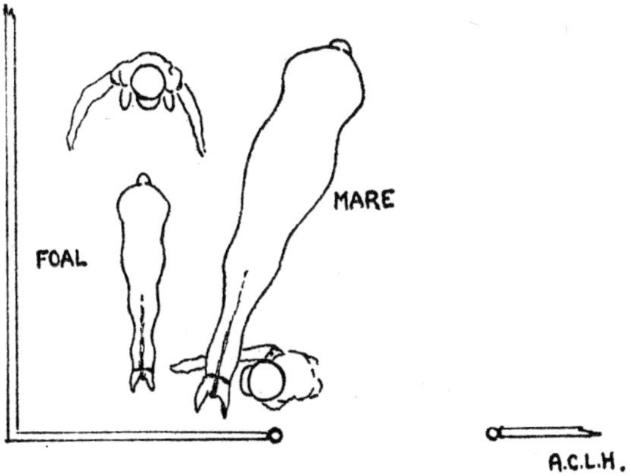

Fig. 3. How to catch a foal in a loose-box

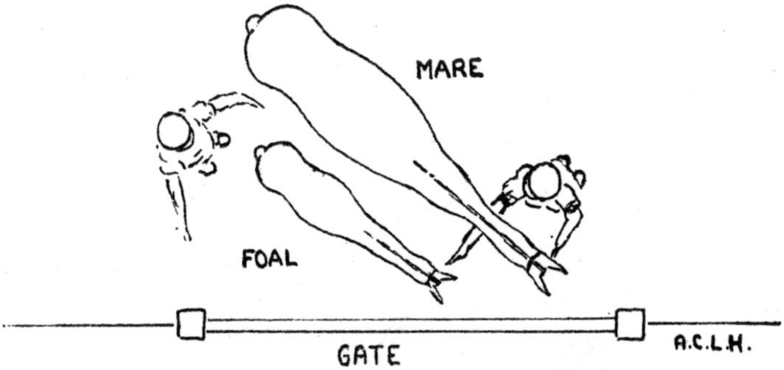

Fig. 4. How to catch a foal in a paddock

find that the foals automatically take up the position shown as soon as the mare is caught. When in the paddock the mare must first be brought up to the gate in order to catch the foal; if you try to catch the foal near the fence, he will probably attempt to jump through as you get hold of him.

When you first teach a foal to lead, it is advisable to take the mare and foal separately, getting one person to lead the mare in front, while another follows with the foal behind. As soon as the foal will lead really well from the head-collar alone, it can be led double with the mare as in Fig. 5.

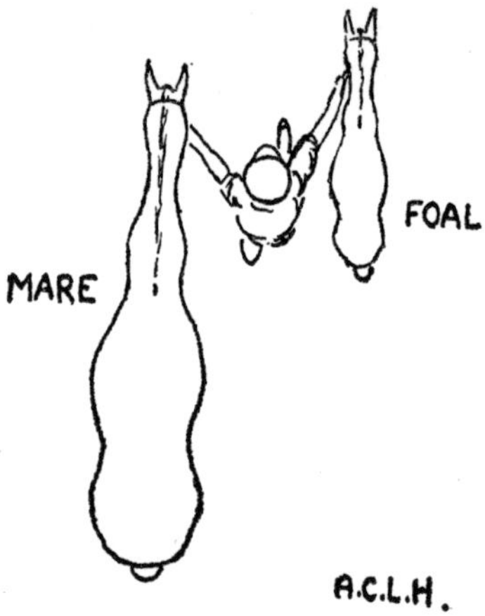

Fig. 5. Always keep the foal ahead of the mare when leading double

Intitially, it is often advisable to get a second person to follow the foal for the first few times, until you are quite sure that it is not going to stop or run back. However, if there is sufficient labour available, foals are better led singly so that they can be schooled to walk out well from an early age.

When leading a mare and foal into a loose-box, always get the foal to walk in first, ahead of you and at arm's length, and allow the mare to follow behind, giving her the full length of the rope if necessary. It is not a good practice to stand outside the loose-box door and let them go; this method tends to teach horses the habit of rushing in and out of doorways. At all costs one should avoid letting horses catch their hips as they go through a doorway. For this reason, therefore, always make a wide turn into a loose-box and, on coming out, walk the horse in a straight line until you are well clear of the door-post before turning.

On no account should people hit foals on their quarters as they let them go into the paddock or into the loose-box. When mares and foals are led out to the paddock separately, the person leading the foal must call out when he is ready to let the foal go; then and not before, the person leading the mare should let her go. You may find yourself in trouble one day, if the foal runs back at the last moment, and the mare has already been loosed.

Very young foals should not be left out for long periods if the weather is wet and cold. Although it may not be raining at the time, they could get a chill if they were to lie down on the damp grass in a cold wind. Once the foals get older and the weather improves, mares and foals should be turned out in the morning after breakfast and need not be brought in until at least 4 p.m. or later.

Many Thoroughbred and Arabian breeders in England now leave their older foals out all night. This is the generally-accepted form of management on most pony studs and has much to recommend it for others. It entails a great saving in labour (sometimes as much as 45 per cent of stud expenses), bedding and hay, which amount to a sizeable annual sum these days. It is, however, essential that all foals are handled, and that they receive a feed each day. It is also vital that any Thoroughbred or Thoroughbred-type mares receive a supplementary ration so that they can milk well and also 'do' the foal they are carrying satisfactorily.

Under this system of management it is advisable to get all the mares and foals in before midday and give them a feed while everything is quiet. While they are in, the owner or stud groom has an admirable opportunity to examine each animal for signs of injury or disease and try any mares where necessary. It also means that a regular worming programme can be carried out and any routine veterinary or blacksmith treatment given. Also, manes and tails can be brushed out each day to keep them tidy, which is very important in the case of show horses, particularly the native breeds and Arabs which are shown unplaited.

It would be inadvisable to bring small pony mares in for a feed, as this type of mare is often very prone to laminitis and is better left out on reasonably bare pasture. When mares are running out all the time it is advisable to construct a creep-feed system in the paddocks, so that the foals can get some extra feed without free competition from their dams. The usual method of doing this is to construct a square fence round some feed troughs. The fence is made with a top rail only, high enough for the foals to get under but too low for the mares to follow. The fencing used must be very strong to withstand being leant on by the mares. It is advisable to put strengtheners on the corner posts and have the posts closer together than normal.

Once foals are eight weeks old, they should be wormed (see Chapter 8). Most veterinary surgeons recommend tubing young foals, to ensure that they receive the necessary dose. Regular worming, at least once every 8 weeks and preferably every 4 weeks, should then be practised, as with all other animals on the stud. It is also absolutely essential that foals' feet are attended to regularly. Once they are over a month old the blacksmith should run a rasp round their hooves. Before this is done the foal must be walked out on level ground and his action checked carefully, so that any deviation from the normal can be corrected before it becomes established. Thereafter foals' feet must be trimmed every month and their action and shape of feet checked regularly for abnormalities.

3

DISEASE CONDITIONS OF FOALS AND YOUNG HORSES

When dealing with the problem of disease, prevention is always better than cure, and is often far cheaper in the long run. So wherever possible I will point out any preventive measures available to owners, as well as the symptoms and when it is wise to call in your veterinary surgeon. A quick guide to the symptoms of the various conditions can be found in Appendix B (page 171).

BARKERS, WANDERERS and DUMMY FOALS

These are terms used to describe obscure conditions sometimes seen in newborn foals. They are attributed to brain damage caused by lack of oxygen at the time of birth, or to infection by the germ *Actinobacillus equuli* or other organisms such as those which cause septicaemia.

Barker foals will often appear perfectly normal at birth but some time afterwards in some cases will start to make a noise like a dog barking, and will show marked signs of respiratory distress and convulsions. If the foal is already lying down, he may jerk his head up and down and move his legs without being able to get up. If he is already on his feet he will probably show symptoms of blindness: wandering around aimlessly, bumping into the walls and continually calling for his mother without apparently being able to recognise her.

The foal should be placed on a blanket to prevent any bedding getting into his eyes which at the very least would irritate them, and at the worst cause blindness. The blanket should be placed under a heat lamp for warmth and an old sweater can be put on the foal. Veterinary aid must be called in immediately before there is permanent brain damage.

Lack of oxygen at the time of birth, associated with premature breaking of the cord, or fractured ribs sustained at the time of birth, can cause these conditions.

Wanderers and dummy foals are variations of the same condition. Wanderers, as the name implies, describes a foal which walks aimlessly round the loose-box, showing definite signs of irritability and restlessness. He will go off suck, chew incessantly or grind his teeth. Dummy foals on the other hand lose consciousness and become comatose.

These foals must be fed by stomach tube if they are to have any chance of recovering. This condition is more common in Thoroughbreds.

DIARRHOEA OR SCOURING

This can be a symptom of several diseases, errors in feeding or worm infestations.

Most foals will suffer from diarrhoea at some period before they are weaned, often when the mare comes in season after foaling. This is thought to be due to changes in the composition of the mare's milk at this time. But the ingestion of poor quality high-bulk feed, such as inferior hay, before the foal's digestive tract is capable of dealing with these fibrous substances, may play a part. The foaling heat scour usually stops quite naturally soon after the mare has stopped being in season. The foal does not go 'off suck' or show any other symptoms of illness but this 'natural' scour can weaken the foal and make it more susceptible to infectious scours at this time.

When foals scour, some Vaseline should be smeared round their hind-quarters, to prevent the hair from coming out. This is especially important if you intend to show or sell the foal before weaning.

All infectious diarrhoeas should have immediate veterinary attention, as young foals have very little reserves and are soon pulled down in condition.

When a foal is scouring there is usually a characteristic odour which is immediately recognisable on entering the loose-box. The foal's temperature is usually raised. The thermometer must be held against the side of the rectum otherwise, in cases where the rectum is relaxed, a false reading might be obtained. The foal may go 'off suck', or may continue sucking. Some foals drink large quantities of water and their stomachs become distended. Foals which stop sucking soon become dehydrated, and in these cases a balanced electrolyte mixture such as Ionalyte should be given by tube. This can also be added to the water of foals that are drinking readily. On no account should water be restricted. Signs of dehydration usually include sunken eyeballs, staring coat and a weak pulse.

Draughts and sudden changes in temperature can be predisposing causes of scouring. When it occurs the most important thing to remember is that any fluid loss must be replaced and the electrolyte balance maintained. If the fluids are not replaced in adequate amounts the foal will die as a direct result of fluid loss. The foal's tongue will often become coated and there is a marked loss in condition and wasting of the muscles.

Injection of some of the mare's blood into the foal will help in some cases due to the antibodies this blood will contain. Antibiotic treatment is usually necessary for recovery.

ENTROPION

This is a condition in which you will notice that the foal has a chronic watering of one or both eyes. This becomes very apparent as early as one or two days after birth. On careful examination one can see that the bottom eyelid has turned inwards and the lashes are irritating the eye.

This is a job for your veterinary surgeon and he may wish to insert some stitches into the bottom lid. This will have the effect of making the lid turn outwards in the normal manner. The stitches are usually left in place for about two weeks.

EQUINE INFECTIOUS ANAEMIA (SWAMP FEVER)

This is a notifiable disease of horses and any suspected case must now be reported to the police or Ministry of Agriculture. It normally occurs in low-lying swamp areas – hence its name – and is transmitted by mosquitoes. It is endemic in the U.S.A. and with the increase in the traffic of horses throughout the world, diseases such as swamp fever are likely to occur almost anywhere. A case was reported for the first time in England in 1975.

This disease can manifest itself in several different ways:

(1) fever associated with anaemia, minute areas of haemorrhage on the undersurface of the tongue and soft swellings on the limbs and other parts of the body, sometimes loss of condition;

(2) chronic loss of condition together with anaemia, ending in the death of the animal;

(3) sudden death without any prior symptoms.

Some affected animals do not show any symptoms unless they are subjected to stress such as foaling. These animals can nevertheless transmit the disease to other healthy stock, given the right conditions.

Diagnosis of the disease is made by taking a blood sample and subjecting it to a special laboratory test known as the Coggins test.

In the United Kingdom no compensation is paid for affected animals, which, if kept, must be isolated for the rest of their lives.

GUTTURAL POUCH TYMPANITIS

This is a very rare condition sometimes seen in foals. The affected animal appears as in Fig. 6, with a large swelling in the region of the guttural pouch. When pressure is applied to the swelling, air is driven out but the swelling returns within a few hours.

Fig. 6. Tympany of the guttural pouch

The affected animal makes a roaring or snoring noise when breathing and often makes more noise when eating. In hot weather the animal may be in some distress. There may or may not be a nasal discarge. This condition might be inherited or it can be due to an infection. Immediately you see the first symptoms you should call in your veterinary surgeon; he may advise an operation. On no account should you attempt to lance the swelling yourself or let any fluid or air out, as this could start up a serious infection.

HAEMOLYTIC DISEASE

This is a blood disease of the newborn foal in which the

chief symptoms are anaemia and jaundice. The disease is caused by a blood group incompatability between the foal and its dam, and is somewhat analogous to Rhesus babies in human medicine.

The condition arises when a foal, while still in the womb, has inherited a blood group from its sire which is incompatible with the dam's blood. The mare then becomes sensitised to this 'foreign' blood group and produce antibodies against it. She concentrates these antibodies in her colostrum so when the foal first sucks they pass into its bloodstream. The antibodies then combine with and destroy the foal's red blood cells, which results in severe anaemia and jaundice.

If you own a mare which has previously had a foal with haemolytic disease or if you want to make sure that your mare does not have a foal with haemolytic disease, then about 3–4 weeks before she is due to foal ask your veterinary surgeon to take a sample of her blood. He will send this away to a laboratory to be tested against samples taken during the succeeding two weeks for an increase in antibody level. Any significant increase in level would denote that the foal will be at risk should it receive any colostrum from its own dam.

If it has been determined that the mare has an increased antibody level and that the foal when it arrives will therefore be at risk, the main preventive measures to take are:

(1) Obtain a foal-size muzzle – this must be put on the foal the moment he gets to his feet, to prevent him from drinking any of his mother's milk and must be left on for 36 hours. If all access to the dam's colostrum is withdrawn the disease cannot occur.

(2) As the foal must have some colostrum an alternative supply must be provided. This can be obtained at any time from another unsensitised foaling mare and kept in a deep-freeze until ready for use. It should be warmed to blood heat before feeding. (See chapter 2, page 28.)

The mare must, of course, be milked out at least six times

a day to prevent her from drying up before the foal is allowed to suck. This milk must be thrown away.

The clinical signs of haemolytic disease are that the foal will be seen to yawn, appears sleepy and short of breath due to lack of oxygen in the blood, from the destruction of the red blood corpuscles. The mucous membranes are yellow. This condition usually becomes apparent within the first 36 hours after birth.

HERNIA

A hernia or rupture is caused by incomplete development of a muscle wall (usually abdominal) and the protrusion of an organ or its fatty connective tissue through the hole which has been formed, causing a swelling to the outside beneath the skin.

The type of hernia most commonly seen in foals is the umbilical hernia. This is seen as a swelling in the navel region and is due to part of the bowel protruding through the umbilical ring, which has failed to close up in the normal way. It usually becomes apparent at about six weeks, gradually increases in size for some time and then starts to get smaller by the time the animal is 9–18 months old.

Umbilical hernia should not be confused with thickening, which sometimes takes place in the region of the navel. This is hard and unyielding, whereas in the case of a hernia a distinct ring can be felt through which the protruding mass can be pushed back into the abdomen, protruding again once the pressure is released.

In severe cases of hernia your veterinary surgeon should be consulted, particularly when there is any possibility of the hernia becoming strangulated, as an operation would be indicated. This also applies to cases where it is intended to sell the foal after weaning, when the presence of a hernia would affect the selling price. The operation is usually carried out when the foal is 3–4 months old, i.e. before it is weaned.

Some people advocate applying blisters to the area, but

this tends to thicken the skin making surgery, if necessary, more difficult. Some hernias disappear without treatment.

INFLUENZA

Affected horses will cough, run a temperature and sometimes have a slight nasal discharge. Adult horses usually recover in about a week, but young foals often become seriously ill and some die, as they have very little resistance to influenza at this early age unless the mare has been vaccinated recently.

The incidence of influenza among foals can be greatly reduced by ensuring that all brood-mares are vaccinated annually and that the booster vaccination is carried out in the last month of pregnancy. This will produce the maximum amount of antibody in the mare's colostrum. No ill effects as a result of vaccination, either to the mare or the foal she is carrying, have ever been recorded. Foals should be vaccinated at two months and again at three months, with a booster dose 12–18 months later.

As far as Thoroughbred yearlings are concerned, it is recommended that vaccination should take place in July or August.

JOINT ILL or NAVEL ILL

This condition usually becomes apparent some time between five days and four months of age, but most commonly at about three weeks. It is characterised by swellings in the leg joints or around the navel. The foal usually goes lame in the affected limb or limbs.

The germs which cause this disease are thought to gain entry at the time of or soon after birth. Maximum cleanliness should therefore be observed at foaling time.

The foal may also go 'off suck' and appear dull and listless, the temperature rising to 103°–105°F. It will tend to spend most of its time lying stretched out flat on its side. One or more leg joints may appear swollen, tense and hot. If left untreated these will eventually burst, discharging blood-stained material, and the foal will gradually get weaker and may die. Sometimes abscesses are also found on the internal organs, in which case death usually occurs.

Prevention mainly consists of maintaining scrupulously clean surroundings at foaling time and a liberal application of antibiotic powder to the cord the moment it breaks. This should be repeated an hour or two later. Make sure the foal receives some colostrum within the first 24 hours of birth, particularly in cases where the mare has been running her milk before foaling. The mare's colostrum and foal's serum can be checked for antibody level. As a final precaution the foal should be given antibiotic cover for three consecutive days after birth.

Should you suspect that your foal has Joint Ill, you should call your veterinary surgeon immediately, as if left untreated, at the very least, there may be some permanent damage to the foal's joints.

MECONIUM COLIC

Meconium is the brown or black fairly hard excretory material which collects in the bowels before birth. It should be passed within a few hours, helped by the purgative action of the colostrum. The meconium is thought to consist of amniotic fluid cells, bile and debris from the intestines which have collected before birth.

It is very important that the meconium should be passed within a few hours of birth to allow normal digestion to take place. If the meconium is retained, colic will occur. The foal will probably carry its tail higher than normal, and may be seen standing in a crouched position with its back arched, straining. In a more advanced stage it will show very definite signs of colic: kicking up at its stomach, rolling and even lying on its back with its legs tucked up against its stomach in an effort to relieve the pain. (See Plates 7 and 8).)

Retained meconium and constipation in the foal may be

due to a costive diet and lack of exercise in the mare.

A dose of liquid paraffin by mouth or an enema of warm soapy water does sometimes help to relieve this condition, but it is probably better to call in your veterinary surgeon as soon as you notice the first symptoms, rather than wait until the foal gets worse. Routine dosing of all newborn foals with liquid paraffin by mouth can help to prevent meconium colic.

This condition appears to be more common in colt foals than fillies, especially overdue colt foals.

MENINGITIS

This is the term used to describe inflammation of the membranes covering the surface of the brain and/or spinal cord.

The affected foal usually goes 'off suck' and appears generally ill, with temperature, respiration and pulse higher than normal. It may become dazed and walk into objects in its path, and develop convulsions.

Veterinary help should be called in immediately as this is a very serious disease and only very careful nursing can effect a cure.

PERVIOUS URACHUS

In the case of this condition a continuous trickle of urine will be seen coming from the navel within the first four days. It will probably take the hair off the lower regions of the hind legs, unless these parts are smeared with Vaseline to protect them.

Before birth the foal lies inside a sac known as the amnion and is immersed in amniotic fluid. This is surrounded by the chorio-allantoic membrane, which contains allantoic fluid, the foetal urine which has collected before birth. The foal's urinary bladder opens directly into the chorio-allantoic sac, which gradually fills with urine, thus preventing over-distension of the bladder. At birth the umbilicus normally closes.

In the case of pervious urachus there is still direct access for the urine through the navel cord to the outside. The condition will sometimes cure itself, but if it does not seem to improve within two weeks your veterinary surgeon should be called as treatment may be necessary.

PNEUMONIA

This is inflammation of the lungs with a resulting breakdown of the lung tissue which greatly impairs respiration. The foal will be observed to be breathing more quickly and heavily than usual. A variety of germs, bacteria and viruses are able to cause pneumonia in susceptible foals.

Predisposing causes of pneumonia may be exposure to cold, wet weather, or stuffy and badly-ventilated stables. It can also be produced by careless drenching – allowing some of the fluid to go down into the lungs, which is always a risk.

Possible symptoms of pneumonia are a high temperature and fast, laboured respiration. The foal will go 'off suck' and there may be fits of shivering and nasal discharge. Careful nursing and management is all-important as a relapse can take place at any time if the standard of care is relaxed.

Veterinary aid must be called in right at the beginning if a good chance of recovery is to be expected.

Summer pneumonia: This is a highly-infectious disease of young foals from about 1–4 months old, caused by the germ *Corynebacterium equi.* As its name implies this condition is most commonly seen during the summer months. It is more commonly found in Scandinavia and Australia but outbreaks do occur from time to time in the British Isles.

The germ causes areas of pus (abscesses) to form in the lungs. By the time symptoms of pneumonia develop the disease is usually fairly well advanced; hence the high death rate.

Antibiotic treatment by your veterinary surgeon and good

nursing give the only possible chance of recovery. Foals with pneumonia of any type should be isolated immediately, as summer pneumonia can spread very rapidly through a stud.

RESPIRATORY INFECTION DUE TO HERPESVIRUS

This virus causes a respiratory infection most commonly seen in yearlings. Under certain conditions it also causes abortions in mares, and occasionally produces signs of paralysis.

In young horses the symptoms are usually those of a common cold: nasal discharge, sometimes accompanied by a slight rise in temperature, reddening of the mucous membrane in the eyes (conjunctivitis) and swelling of the glands in the angle of the jaw.

It usually clears up on its own, but where there is a rise in temperature, however slight, your veterinary surgeon should be consulted. (See also Snotty noses, page 51.)

ROUND JOINTS

These are the result of inflammation affecting the growing plates (epiphyses) above the joints of young horses and foals. They are usually caused by trauma or errors in diet such as an excess of phosphorus and too little calcium in the ration, or lack of vitamin D (the sunlight vitamin). They can occur when a high-protein ration is being fed, without a corresponding increase in the calcium level of the diet, particularly in heavy-topped animals, when the ground is hard.

The swellings are sometimes hot and tender and the affected animal will go lame. In other cases they will appear firm and cold. 'Knobbly fetlocks' are quite commonly seen in Thoroughbred foals between 3–6 months of age, and they usually come right in time. (See Fig. 7a.) Swellings involving the knee or hock joints are more common in yearlings. (See Fig. 7b.)

Treatment usually consists of correcting the diet. Your veterinary surgeon may wish to take blood samples to deter-

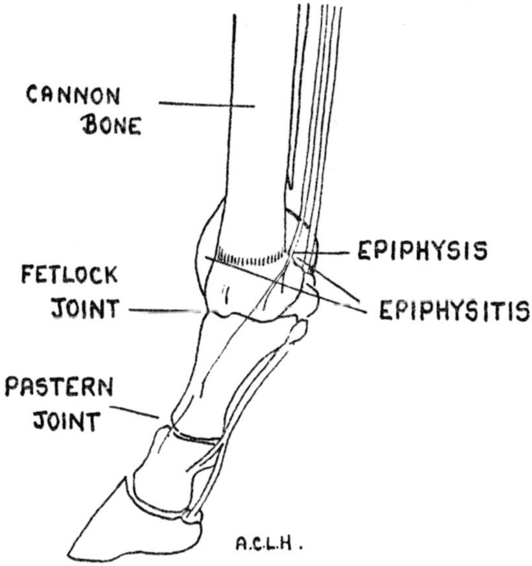

Fig. 7a. Round joints (fetlock)

mine if any gross calcium/phosphorus imbalance is present. Exercise should be restricted and affected animals must be put on soft going, say in a barn.

One should not feed bran to affected horses, as this is very high in phosphates. Calcium carbonate or steamed bone flour should be added to the diet and the protein level of the ration reduced.

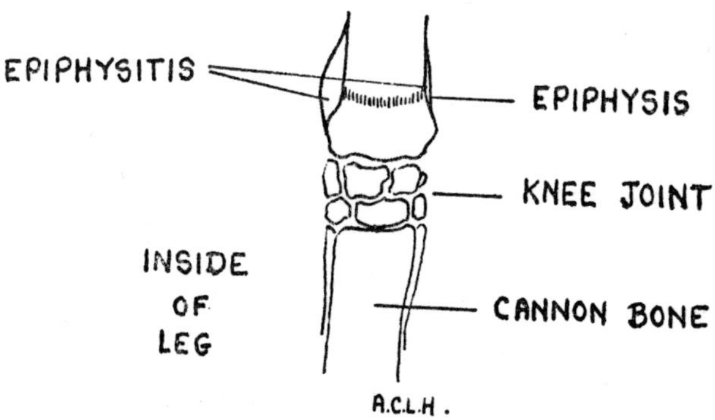

Fig. 7b. Round joints (knee)

Horses which are in show condition or being prepared for sale, i.e. 'heavy-topped' animals, are most prone to this condition. Their leg joints must be watched very carefully and their diet and management adjusted accordingly if round joints are to be avoided.

RUPTURED BLADDER

This is a rare condition, probably due to some inherited weakness in the bladder wall.

The symptoms are similar to those of meconium colic (see page 45) and usually become apparent around the third day. The foal's stomach becomes distended with fluid and it may only be able to pass very small quantities of urine.

Your veterinary surgeon should be called in immediately this condition is suspected, as an operation will be necessary to repair the tear in the bladder wall.

SALMONELLA

As with calves, if not treated with antibiotics from the outset this is a highly infectious killer disease of foals. It causes epidemics of scouring on studs.

The symptoms usually consist of a watery, bloodstained diarrhoea, very high temperature, colic and rapid breathing.

Veterinary aid must be called in immediately this disease is suspected, and any affected foals isolated. Loose-boxes which have housed sick foals must be very thoroughly disinfected, under the guidance of your veterinary surgeon, before any further foals are put into them.

SEPTICAEMIA

In this disease, bacteria have somehow managed to gain entry into the foal's bloodstream, where they circulate round the body and invade any or all of the vital organs. A predisposing cause of this disease can be lack of colostrum and thus the antibodies against disease which it contains. In most cases the affected animal dies, the course of the disease is often very short. It usually occurs within the first four days of life.

Sometimes the only symptoms noticed are a high temperature and general signs of illness followed soon afterwards by death. The sooner a veterinary surgeon can be called in the more chance he will have of saving the animal's life.

SHAKER FOALS

This is a condition sometimes encountered in young foals 3–8 weeks old. The disease appears to be peculiar to America – only one case has been reported in England to date.

The condition is rapid in onset. The affected foal is usually found lying on its side, unable to rise without assistance. When helped to its feet, it will usually start to suck normally but muscular trembling (shaking) soon occurs and the foal will drop to the ground. Death usually follows in about 72 hours. The exact cause of this syndrome is obscure at the moment but veterinary aid should be called the moment the condition is suspected.

SLEEPY FOAL DISEASE

This disease is caused by the germ *Actinobacillus equuli*, and usually becomes apparent 2–4 days from birth.

The affected foal usually appears dull and may go 'off suck'; the temperature may be sub-normal. After several hours the foal appears sleepy – hence the name – may show signs of colic and diarrhoea or develop convulsions.

This organism usually attacks the kidneys and brain.

The sooner veterinary aid is called the greater is the foal's chance of survival.

SNOTTY NOSES

This is usually caused by a virus infection accompanied by a secondary bacterial infection. Affected horses have a runny nose, sometimes cough and run a slight temperature. Their glands may be enlarged.

It usually clears up on its own, but a veterinary surgeon should be called if there is any rise in temperature.

A predisposing cause is badly-ventilated stables. Horses with snotty noses should not be worked until the condition has cleared up.

STRANGLES

This is a disease more commonly seen in young horses up to five years old. It is caused by the germ *Streptococcus equi,* is very highly contagious and will cause the death of young foals. Incubation is usually from 2–14 days. Infected horses show a rise in temperature, appear dull and often go off their feed. They usually have a cough, sore throat, profuse discharge from their nostrils and swellings in the angle of the jaw. The latter causes the horse to hold its head and neck out stiffly. After a few days an abscess will form in the swelling, mature and finally burst, discharging bloodstained pus. This pus and the nasal discharge are both highly infectious, so very great care should be taken that anyone handling an infected horse does not come in contact with other unaffected horses at this time. As soon as the abscess has burst the temperature usually becomes normal and the swelling subsides. Recovery is usually complete in a few weeks.

Treatment consists of immediate isolation of the affected animal and calling in your veterinary surgeon.

TETANUS (LOCKJAW)

This disease affects horses of all ages but its incidence among young foals can be reduced to a very great extent by giving a mare which has already had primary tetanus toxoid injections a booster injection about three weeks before foaling. She will then produce antibodies which will be transferred to the foal in the colostrum, giving it immunity to tetanus for the first few weeks of its life. Immunity can be maintained up to about three months of age by giving an injection of tetanus anti-serum at about six weeks.

This disease is caused by the germ *Clostridium tetani*, which can survive for long periods in a dormant state outside the human or animal body. It cannot live in the presence of atmospheric oxygen. This germ normally gains entry through deep wounds, such as those in the feet, any deep puncture wounds, or through the navel at birth, i.e. any site not immediately in contact with the atmosphere.

The germs do not leave the site of entry but produce a toxin which they release into the bloodstream. This acts on the central nervous system, producing symptoms of paralysis.

The horse shows unusual signs of nervousness. He will stand with his limbs stiff, tail raised stiffly, head and neck stretched out and ears pricked. If the head is raised quickly the third eyelid (haw) will flick across the eye very noticeably.

If tetanus is suspected your veterinary surgeon must be called immediately. But the principle that prevention is better than cure is most certainly true where tetanus is concerned. The preventive measures available are:

Tetanus anti-serum – this is fast acting but only provides immunity for about four weeks or less. It has fewer side effects than tetanus toxoid (*see below*) and is consequently invaluable for very young foals. It is also used after accidents and operations to give the animal temporary passive immunity to this disease while it is most at risk.

Tetanus toxoid – is slower acting and is used to give lasting, active immunity. It is given at three months, with a second injection one month later. This is followed by a booster injection a year after the last injection to give life-long immunity. Where an animal is at particular risk, or in the case of broodmares, annual vaccination is recommended to boost the antibody level.

WARTS (MILK WARTS OR GRASS WARTS)

These warts appear as clusters of small skin-coloured lumps on the muzzle and lips of young horses. They are commonly seen in yearlings at grass during the summer – hence the name 'grass warts'.

They usually disappear on their own after about 2–3 months without any treatment, as soon as the affected animal has developed an immunity to the virus infection which causes the warts to form.

WEAK OR DEFORMED LEGS

Foals, like babies, can be born with various weaknesses and deformities of their limbs (see Plate 6). Some of the more commonly seen are:

(1) *Over-long weak pasterns:* in these cases the foal walks with his fetlocks touching the ground. This condition usually rights itself as the foal becomes stronger. In most of the milder cases of limb weakness in young foals, exercise out in the paddock will help better than anything to strengthen their legs.

Since the fetlock is touching the ground it will become very sore if protective bandages are not applied. Care should be taken that the joint does not become infected, otherwise an infective arthritis (Joint Ill) might result. Once the fetlock has come up off the ground the bandages may be left off, but any sore places should be treated with an antibiotic spray, some of which also have the effect of hardening the skin.

(2) *Contracted tendons:* the foal may only be able to walk on the tips of his toes, or in very bad cases be entirely unable to get up. If the contraction is very bad the foal may have to be put down, but in some cases an operation to cut the tendons and straighten the leg may be successful. In other cases splinting the leg may work well.

When foals are born with twisted legs which do not start to straighten within a week of birth, or where their legs start to twist soon after birth, your veterinary surgeon should be consulted, as the longer these weaknesses are left the harder they are to correct. Correction must take place while the foal is still growing, and may take the form of plaster casts. In some cases a leg which is growing unevenly may need to be stapled to straighten it but this is a major operation. (See Fig. 8.) Feet which turn in or out can usually be

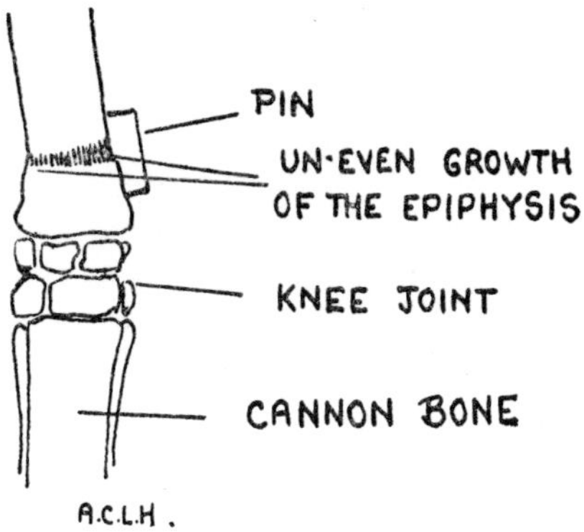

Fig. 8. Weak or deformed legs

corrected quite satisfactorily by the blacksmith. Any minor deviation can usually be rectified by careful trimming but any major fault will probably need special shoeing from the moment the foal's hoof can take a shoe (usually eight weeks old, although shoes which are stuck on may be applied earlier). The outside of the shoe is built up to turn the toe out, and the inside built up to bring the toe in. More slip can be achieved if the front of the shoe, at least, is flattened. Corrections to any joint must be made gradually, otherwise an enormous strain will be placed on the affected limb.

WOBBLERS

This is a condition more commonly seen in colts than fillies and is usually first noticed when the animal is under two years old. Typically the affected animal loses normal control of his hind legs, which become inco-ordinated. This is most marked if the animal is trotted out fast and then brought to a sudden halt, when he will sway and almost fall over behind. Many wobblers are incapable of backing or turning in a tight circle.

This condition never gets any better. Severely affected wobblers should therefore be destroyed, as they are all unsafe to ride.

Horses sometimes become wobblers after an accident, such as rearing up and coming over backwards, or getting cast in their boxes.

The condition is due to changes in the spinal column which adversely affect the cord, which in turn produces the incoordination of the hind-quarters.

REFERENCE AND FURTHER READING:

Rossdale and Ricketts, *The Practice of Equine Stud Medicine*, Baillière and Tindall (London) (see pages 194–318).

4

THE ORPHAN FOAL

An orphan foal is strictly speaking a foal which has lost its mother, fortunately a rare occurrence. But foals may be deprived of their mother's milk for a variety of reasons, for example the mare rejecting her foal at birth, foaling down without any milk, or the foal becoming ill, going 'off suck', and the mare not being milked out regularly enough to keep her in milk. Any of these occurrences will mean that the foal must be reared by hand or a suitable foster mother found.

Rearing foals by hand

The most important thing to remember is that the foal *must have colostrum* within the first 24 hours of birth, so that it can develop resistance to most of the diseases with which it is likely to come in contact (see Chapter 2, page 27). All breeders should, as far as possible, keep a supply of colostrum in a deep-freeze, so that some is available instantly in case of need. This must only be taken from mares which have not produced a haemolytic foal (see page 41) or run their milk before foaling. Deep-frozen colostrum will keep for almost a year. Colostrum from donkeys can be given in the absence of mare's colostrum, but cow's colostrum is not recommended, as apart from the difference in composition between the two milks, it is unlikely, unless the cow in question has been living in close proximity to horses, that her colostrum contains any of the antibodies which will give the foal resistance to normal equine diseases.

If it is absolutely impossible to obtain any colostrum in time, your veterinary surgeon *must* be called, so that he can give the foal some substitute colostrum and antibiotic or serum injections.

All foals must have some form of milk in order to survive. For the first six weeks the foal must be fed every two hours both day and night. Then the length of time between feeds can be gradually increased: every three hours after six weeks, increasing to every four hours at 12 weeks. Foals cannot be treated like calves, but must be fed little and often: overfeeding at any one time will make a foal ill.

At first the foal should be fed from a bottle fitted with a rubber lamb's teat, obtainable from most agricultural chemists, and measuring about 4 in. by 1 in. Milk is also an ideal food for bacteria (germs), so any bottles, teats and mixing utensils used for feeding must be kept absolutely clean and sterilised between each feed. Sterilisation can be carried out with boiling water or by using a dairy chemical or nursery sterilising agent, such as hypochlorite, in which case care should be taken to make sure that the surfaces to be sterilised are absolutely clean before treatment.

If the equipment is not sterilised after each feed, the foal will probably start to scour and become quite ill. To avoid digestive upsets the milk should always be fed at the same temperature, preferably blood heat. To achieve this it is often better if the same person feeds the foal all the time wherever possible. Unused milk should be thrown away and never re-warmed. Milk should not be prepared and left exposed to the atmosphere for any length of time before feeding, as it easily becomes contaminated.

A satisfactory substitute for mares' milk can be made up easily in the following way:

2 tablespoonfuls of lime-water, *not* lime juice cordial, (obtainable from any chemist)
4 tablespoonfuls of glucose
made up to 1 pint with warm cows' milk.

Goats' milk, when it is available, will give better results than cows' milk and should be used as a direct substitute for cows' milk in the above recipe.

An alternative method of rearing foals by hand is to use Ostermilk No 2 at the following rates:

From birth to about 1 month, or according to size of foal: (per feed)	*From about 1 month to weaning:* (per feed)
4 measures of Ostermilk and	10 measures of Ostermilk and
2 measures of glucose to	4 measures of glucose to
8 fluid oz. water	1 pint of water

The Ostermilk should be mixed to a paste with a little cold water until there are no lumps left, then hot water added and the temperature adjusted to blood heat. Too much liquid will cause the foal to develop a pot-belly.

Should the foal develop a digestive scour, as with bucket-reared calves the milk substitute should be withdrawn and replaced by glucose and water only, for a day or two.

Each feed should measure from 15–25 fl. oz. or more according to the size of foal. Ostermilk is really designed for human babies and is a rather expensive method of rearing foals by hand through to weaning. There is a milk powder on the market formulated exactly to the requirements of orphan foals, known as Equilac. This can be obtained from the manufacturers Pegus (Pegasus) Ltd. It contains all the vitamins, minerals and trace elements necessary for a young foal, in a high-quality milk powder composition and should be fed according to the makers' instructions (there is no need to add glucose or lime-water):

Age	Equilac	Water	Pegus (Pegasus) Junior nuts
0–2 days	Colostrum (Biestings)		
3rd day	1 lb.	4 pints	
4th day	1½ lb.	6 pints	
5–7 days	2 lb.	8 pints	
2–4 weeks	3 lb.	12 pints	½ lb.
5–9 weeks	4 lb.	12 pints	½ lb.
10–12 weeks	4 lb.	10 pints	2 lb.
13–14 weeks	2 lb.	5 pints	4 lb.
15–16 weeks	1½ lb.	4 pints	6 lb.

Most foals are born with a well-developed suck reflex and those which are slow in this respect will usually develop the desire to suck soon after getting to their feet. Ideally, the foal should have its first feed within two hours of birth and certainly not later than four hours. Warm up some colostrum or if none is available call your veterinary surgeon. Otherwise give a feed of substitute mares' milk as described above. If you have no milk or equipment, it is better to wait until the shops open in the morning than risk upsetting the foal with a home-made substitute. In cases where the mare dies foaling, some colostrum can usually be drawn out of her udder just before she dies, which can then be given to the foal.

To get a foal to drink from a bottle, first make sure that the hole in the rubber teat is large enough to allow a good flow of milk to escape. Give the foal a taste of the milk by squirting a little into its mouth – the teat can be smeared with a little honey. This should awaken its suck reflex. If this fails, make sure your hands are clean, run a little milk on to your fingers and get the foal to suck them, then gradually substitute the teat for your finger. Unless he is lying down, it is better to get the foal's hind-quarters into a corner, so that you will have some control over him. If the foal does not start sucking soon, squeeze some milk out of the end of the teat into his mouth, until he finally gets the idea. When the foal is hungry he will soon start to suck without much trouble.

The manufacturers of Equilac also make a Junior nut which they recommend should be fed with the milk substitute. It is best if the foal can be encouraged to eat some form of solid food as soon as possible. Most foals will nibble a little warm bran mash from your fingers the day after they are born; something warm and wet will encourage them more than a dry feed. Therefore a little bran, boiled linseed, rolled oats and flaked maize is highly suitable. Once the foal is beginning to eat, he should be introduced to the Junior nuts as suggested in the table; after weaning he should be fed at the rate of 1 lb. per day per month of age.

Ad lib feeding of nuts is perhaps to be preferred, as the foal can then help itself whenever it wishes. This is a particularly useful method for all foals in cases where the owner goes out to work each day, in which case a creep feeder can be used. A boiled barley and linseed mash containing a raw egg can be fed every night, to give some variety to the diet. The foal must also receive *ad lib* hay and water.

A foal which is to be reared by hand must be put into a clean loose-box – one which has not had a sick, and in particular a scouring foal in it that season. As far as possible it should be provided with a companion, otherwise it will become very lonely and will tend not to thrive. A quiet sheep, goat or big dog is suitable. Make sure the foal is kept warm. In cold weather an infra-red lamp may be necessary, or a rug can be made from a woollen jumper, putting the foal's front legs through the arms.

Rearing the foal on a foster mother

Most hand reared foals become very 'humanised' and seldom have much respect for their owners. It is therefore better, if at all possible, to bring an orphan foal up on a foster mother.

If you intend to find a foster mother for your orphan foal, do not teach it to drink out of a bowl, as it then tends to loose the suck reflex and it is impossible to get it to suckle a mare later on. (If you do wish to teach the foal to drink out of a bowl or bucket, the easiest way is to get him to suck your fingers and then gradually lower your hand into the milk, in exactly the same way as one would for a calf. A little food scattered in the bottom of the bucket when the foal has nearly finished drinking will encourage him to start eating.)

Rearing a foal by hand is a tedious but rewarding job, necessitating getting up every two hours throughout the night for at least six weeks. The problem of finding a suitable foster mother without any help, is usually insurmountable but there is an organisation in England, known as the National Foaling Bank, which is geared to help anyone who has either lost a mare or foal. It has handled over 5,000 cases in 10 seasons. The address is:

Miss Johanna Vardon,
Meretown Stud, Newport, Shropshire.
Telephone number: Newport (Salop.) 811234 (STD code 0952).

A suitable foster mother is a mare which has suckled a foal before and is the right size – small pony mares are unsuitable for large Thoroughbred or hunter foals – and has plenty of milk. If you are lucky enough to find one, then the problem is to persuade her to accept the orphan foal. Some mares have a very well-developed maternal instinct and will readily accept any foal; others are not quite so keen. In either case one must take certain steps to make the mare believe that the foal she is being asked to foster is her own. Intially mares learn to recognise their own foals by the smell of the amniotic or birth fluids. Therefore this smell must somehow be transferred to the orphan foal. Later the mare learns to recognise her foal by sight. The mare can be deceived in two ways:

(1) If the after-birth is still available and quite fresh, the orphan foal should be rubbed all over with it. Make quite certain that you include his head, legs and tummy, then spread the after-birth out over his back, as you would a rug. Push him into the mare's loose-box, keeping hold of him and making sure someone is also holding the mare. Allow the mare complete freedom of her rope, so that she can examine and smell the foal, but watch the expression on her face carefully and be prepared to move the foal out of harm's way very quickly, should she show signs of rejecting it. If the mare seems to be happy with her new baby, you can then let him go to the udder for a drink. If the mare is still happy, the after-birth can be removed and the mare and foal left alone, although a careful watch should be kept on them for the next few hours, just to make sure that the mare does not change her mind.

(2) Another method is to skin the mare's own dead foal, put the skin over the back and legs of the orphan foal and tie it in position like a rug with pieces of string. However, you must make absolutely certain first that the foal has not died of any known contagious disease, such as scouring, otherwise the healthy foal might very well die of the same disorder. Leave the 'rug' in place for 48 hours and then take it off. Do not leave it on the foal so long that it starts to go bad – which does not take very long where heat lamps are used in a foaling box. Strong-smelling ointment smeared into the nostrils of mares whose foal was a few days or weeks old when it died will help to camouflage the smell of the orphan foal. As a safety precaution the orphan foal can be left in a solid partitioned section of the mare's loose-box so that she can see but not reach the foal to injure it and the two only allowed together at feeding time, when the foal should wear his 'rug'. This is done until such time as the mare shows obvious signs of accepting the orphan foal. Fractious mares should be doped to make them more tractable, on the advice of a veterinary surgeon.

Mares offered as foster mothers must be in good health and have plenty of milk. Foals offered for fostering must also be in good health and not 'on their last legs'.

5

WEANING

Weaning is the process of separating the mare and foal when the foal is old enough to thrive on a solid diet alone.

The very earliest age at which this can take place is when the foal is four months old. The optimum time for the average mare and foal to be weaned is at six months old, but the exact time choosen will to a very large extent depend on the general condition and health of both mare and foal, the size and maturity of the foal and also whether the mare is pregnant or barren.

If the mare is in-foal again, even though she may be fat, it would be far better, in the interest of the foetus she is carrying, to wean her present foal by six months. In the case of an in-foal mare in poor condition, it would be definitely advisable to wean the foal as early as possible – at four months. Fat barren mares, on the other hand, can be left with their foals until the following spring should the owner so desire.

If the foal at foot is in any way weak or ill, it must on no account be weaned until it has recovered, as the act of weaning is always a shock to the foal's system, although the degree of shock depends to some extent on the method used. Weaning would undoubtedly set the foal back further. In cases where a sick foal goes 'off suck', the mare must be milked right out at least six times a day, to keep her in milk until the foal has recovered and starts sucking normally again.

Foals must be prepared for weaning if a major set-back is to be avoided. This means that they must be introduced to the exact solid diet they are going to receive after weaning, some three to four weeks beforehand. In this way they are eating their food readily by the time they are weaned. This is most important, otherwise the weaned foal will loose condition rapidly during the first few weeks of separation. The easiest way to get the foal to eat solid food readily is to feed both the mare and foal, spreading the food out in a long trough or manger, so that both can eat without too much free competition. The foal will soon imitate its mother and develop a liking for the food.

As foals get older they tend to spend less and less time close to their mothers and more and more time playing with other foals in their group. They will establish friendships between themselves and this factor can be utilised to advantage where weaning is concerned.

Basically there are three methods of weaning foals:

(1) *Paddock weaning*

This method has the advantage that it can be used where there are foals of different ages on the stud. Whenever it is possible this is definitely the method of choice, as it produces minimum set-back in the weaned foals as compared with the other methods described below. Also the foals need not be stabled but can be left out day and night if preferred.

Prior to weaning, the mares and foals are collected together into the same paddock so that the foals can get to know one another and become accustomed to their surroundings. At least two weeks should be allowed for this. It does not matter whether the mares and foals are stabled at night or if they are running out all the time. Before weaning starts check all the fences, making quite certain that they are absolutely stock proof. Select a second paddock or field, out of earshot of the foals' paddock, into which you can put the mares after weaning – make sure this paddock is also stock proof.

Weaning should always take place in the morning and on days when there are going to be plenty of people about to keep an eye on both the newly-weaned foals and the mares. Before weaning any of the foals, a plan of campaign must be worked out. Any vicious or restless mares should be removed first. On the initial day two mares should be taken out so that they can be turned together into their selected paddock. After this the mares can be removed one at a time, if so desired, the quietest and most gentle mare being left to last. No more than two mares should ever be weaned at the same time, otherwise there will be too many foals at once, looking for their mothers in the same paddock, which will defeat the object of the exercise.

On the selected morning, if the mares and foals are running out at grass all the time, the two mares to be removed are caught up and led off to their distant paddock, and the weaned foals left behind with their friends. If on the other hand the mares and foals are stabled at night, then the mares and foals are turned out as usual, the two foals to be weaned being left to last. They are led out of their boxes and up to the paddock gate as usual, but at this point the foals are led into the paddock to join their friends and the mares are led away. The foals should not be loosed until the mares are right out of sight. A third person should walk with the mares, to chase them up, should they try to stop and a watch must be kept on both the mares and foals until they have settled down.

The two weaned foals will normally gallop up and down the fence a few times before going to join their friends and start grazing and playing as usual. They never go to any of the other mares and normally settle down completely within an hour or two. In a matter of about three days another mare can be removed and so on until only one mare remains.

The last mare can be left with the weaned foals for several weeks if so desired or she can be removed after a few days. (See Plate 9.)

It is better not to put too large a group of mares and foals together for weaning. About ten mares is the maximum number for ease of operation, although no real danger is likely to occur if the number is greater.

Where the mares and foals are normally stabled at night the weaned foals should be brought in as usual and they *must be put back into their usual loose-boxes,* so there is the minimum of change in their daily routine. It is also absolutely essential that the ration which has been given before weaning is maintained after weaning, if a major set-back is to be avoided.

The mares will not require any special treatment after weaning. Unless the weather is very wet and cold, they should be left out in their paddock day and night – this will help them walk off the excess milk in their udders. If they are brought in at night they must only receive hay and water to encourage them to dry up quickly. On no account should any milk be drawn off, as this will only cause more to be formed and delay drying-up, and will also tend to cause mastitis. Nature works on a supply and demand basis. If the demand for milk ceases then she will make certain that the supply also ceases.

Signs of mastitis in mares are usually a hot swollen udder with swellings extending forward along the underside of the belly and between the hind legs. The affected mare will usually walk stiffly and often go off her feed. If a little milk is drawn from the udder it will be seen to be thick and lumpy. Your veterinary surgeon should be consulted as he may wish to prescribe antibiotic treatment. Untreated mastitis can result in the loss of use of the affected parts of the udder.

(2) *The traditional method*

Under this system the weaned foal is put into a roomy loose-box and the mare is turned out into a well-fenced paddock right out of earshot of her foal, as described above.

The loose-box to be used for the foal should be large enough to allow him to have some exercise and not get cast. The water bucket should be placed off the ground and any hay nets, if used, should be removed, at least until the foal has settled down. If replaced they should be tied up as high as possible, as young horses tend to play around and can very easily get a foot caught in a loose hay-net. The manger system as described on page 80 is probably the safest and most practical for all horses.

Plenty of bedding should be put in the loose-box to prevent injury. The droppings can then be picked up each day, but it will not be absolutely necessary to muck the box right out until the foal is turned out again in a few days' time.

As weaning usually takes place in the late summer or early autumn when the weather may be hot, a wire cage should be fitted over the top door of the loose-box so that the foal can get plenty of air and ventilation without being able to jump out. If your loose-box is not fitted with a wire cage, one can be improvised quite simply by making a rough wooden frame the same size as the top door of the loose-box and nailing some ½ in. wire netting over the frame. This can then be tied in place with string to the hinges and fittings of the top door.

On no account should foals be shut away in darkened loose-boxes at weaning. Some people shut the top doors of the foals' loose-boxes and cover the windows up, maintaining that the foals settle down better if they are kept in semi-darkness. This is probably the finest way of encouraging foals to develop such bad habits as crib-biting and weaving, and is not to be recommended.

On the day selected for weaning bring the mare and foal in from the paddock and put them in the prepared loose-box. Put a bridle on the mare. With the help of at least one person, walk the mare out of the box, leaving the foal alone inside and shut *both* top and bottom doors. Make sure someone is walking behind the mare to prevent her from stopping or trying to run back to her foal as you take her to the paddock.

Several mares can be weaned together and turned out into the paddock at the same time. If only one mare is to be weaned then a suitably quiet companion must be found to run with her, at least for the first week or so until she has settled down completely.

Turn the weaned mares loose in the paddock and stand by the gate until they settle down and start to graze peacefully, which may take an hour or more. Once the mares have settled it is quite all right to return to the yard and see to the weaned foals. If it is a hot day the top door of the loose-box should be left open with the wire cage in place. The foals can be given a feed – this *must be exactly the same ration they were receiving before weaning.*

After about 2–3 days the foals can be turned out for exercise into a well-fenced paddock where they can easily be observed. If only one foal has been weaned a very quiet companion, preferably of about the same size, should be found to run with the foal. If no such companion can be found then the foal should only be turned out for short periods at a time, when he can be watched carefully, as foals are completely without any fear or knowledge of their own capabilities. This can prove pretty lethal at times, as they will often take on fences, hedges and gates which would stop any normal self-respecting adult horse.

Weaned foals must be brought in at night for at least 2–3 weeks after weaning. If you do decide to turn them out again a dry warm night must be chosen for the first night out, otherwise the young horses could easily get a chill.

(3) *Weaning by pairs* No They could become inseparable

Some people prefer to wean two foals at a time and put them together in the same loose-box. This method has the advantage that foals tend to settle down quicker than they do under the traditional method. However it does entail a further period of weaning at some later date when the two foals have to be separated. Another disadvantage is that one foal is almost certainly going to be 'boss', so unless the food is supplied *ad lib*, or a constant watch kept at feeding time, one is bound to get more than the other.

Another method used on some studs is to put a metal grille in the dividing walls of the loose-boxes, so that the horses can see their neighbours at all times.

When horses are left entirely to nature, weaning normally takes place when the foal is about a year old, shortly before the next foal is born. This is not practical where horses are being bred for most domestic purposes, such as showing or the sales. In these cases the foals need to be separated from their mothers so that they can be handled and fed in preparation for their future role in life, and the shows or sales in which they are to be entered.

Some colt foals will be seen to mount their mothers as early as one month old, but they do not usually 'draw' until they are about three months old. Colts are not sexually mature enough to get a mare in foal until they are about 16 months old, but they should be separated from the filly foals at about nine months, as their play is so much rougher than the fillies' that accidents can happen only too easily.

6

FEEDING YOUNGSTOCK

The old saying about all roads leading to Rome is never more true than when it comes to feeding horses. Most established breeders have well-tried methods which they find work best on their own studs; so I propose to describe my own particular 'road to Rome' and let the reader decide which road he or she prefers to take.

The type of land and climate is the biggest single factor which must be taken into account when devising a method of rearing young horses. Where the land is poor and the grass tends to burn up during dry spells one has to use expensive concentrates to achieve the same results as the breeder who is on good land in a mild area with a relatively high rainfall.

When breeding horses or ponies specifically for the sales or show-ring, one aims at producing a well-grown precocious type of animal which will stand out among his contemporaries and catch the judge's or buyer's eye instantly. This is often difficult to achieve without running into trouble (enlargements in the region of the knee, hock and fetlock joints (see Fig. 7)) unless the breeder is prepared to regulate the amount of flesh the young horse is carrying with the skeletal development and type of bone of each individual.

A knowledge of the science of feeding is essential so that any ration can be brought within certain limits which have been worked out to give optimum results under average conditions. For the convenience of breeders who have neither the time nor the inclination to devise their own rations some animal feed manufacturers market a wide range of balanced cubes for every stage and condition in a horse's life, from the foal, through weaning, to its emergence as a racehorse, hunter, child's pony, etc. All the feeder needs to do is obtain the necessary feeding instructions from the manufacturer and follow them to the letter.

For the benefit of those breeders who like to make up and feed their own rations, either for reasons of economy or pleasure, I will go into the subject more deeply.

A basic understanding of the digestive tract in the horse is essential to good feeding (see Fig. 9). The rate at which any feed travels along this tract governs its digestibility and degree of assimilation. It has been estimated that, on average, it takes about 72 hours for any feed to clear the digestive tract completely; but most food will pass through the stomach and small intestines in under 12 hours. The latter is the area in which most of the protein, carbohydrates, fats

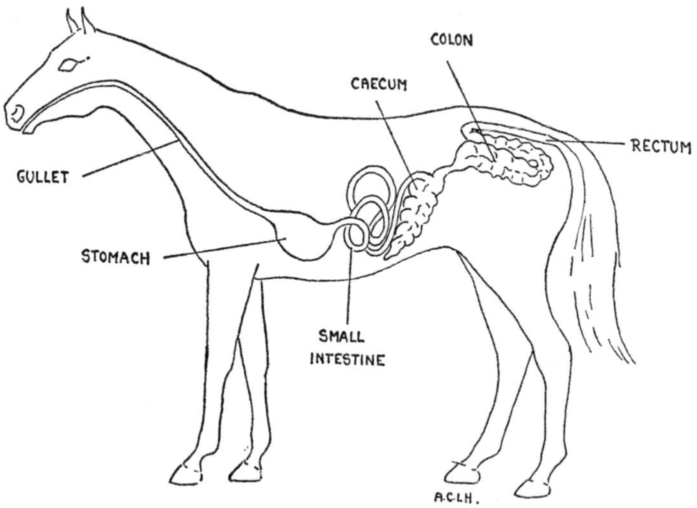

Fig. 9. Digestive system

and minerals are absorbed.

It follows, therefore, that the larger the quantity of food passing through these areas at any one time the lower the degree of actual assimilation; so it is better to feed horses little and often. The large intestine contains bacteria which break down the fibre content of the ration into volatile fatty acids which are then absorbed. Any protein which has not been broken down into its constituent amino acids and absorbed from the small intestine enters the large intestine and is utilised by the bacteria for reproduction. These in turn become available to the horse as a source of protein or are passed out in the faeces. It is known that the very young horse, which has no microbes in its gut, is incapable of utilising non-protein nitrogen (urea). In which case the feeding of urea nuts to young horses is only a waste of time and money. As the microbes live in the large intestine and most of the protein is digested in the small intestine, it therefore follows that the horse should be fed only high-quality proteins in its diet.

Protein is composed of numbers of different amino acids and is essential for building tissue. An increase in the protein fraction of the diet is therefore necessary for young growing stock and pregnant mares in late gestation, but not for working horses who only require an increase in the energy value of the ration. Care should, however, be taken not to overfeed protein. Protein requirements (on a dry weight basis) have been estimated as follows:

Foals	20% protein in their diet
Yearlings	14% protein in their diet
Two- and three-year-olds	10–12% protein in their diet

The protein constituents of the ration must be of the best quality and should be drawn from a wide variety of ingredients to give the greatest range of amino acids in the diet.

As long as it is being fed on a natural diet it is impossible for a young horse to suffer from an actual protein deficiency.

Only very fast-growing species (e.g. rats) will normally show marked signs of protein deficiency. However, young horses given inadequate levels of protein will not do so well as those on optimum diets.

As we have just stated, protein is broken down into amino acids and then re-built as animal protein. Any unwanted amino acids are discarded in the urine and faeces. Some amino acids are more important than others, several being absolutely essential to the maintenance of life, examples of which are lysine and methionine. These are found in spring grass and other high-quality protein feeds, such as soya bean meal. They are not found to the same extent in good-quality hay, oats and other cereals. In cases where high-quality protein feeds are not available, better growth results may be obtained by feeding supplements containing these limiting or essential amino acids.

A foal will not consume very much solid food for the first few weeks of its life, but will rely on its mother's milk. During this period of maximum growth it has been estimated that the foal requires just over $\frac{1}{2}$ lb. of digestible protein for every 1 lb. gain in body weight, which is supplied initially by the milk alone. The young foal has not developed a natural bacterial-flora in its large intestine by the time it starts to eat, and therefore it should only have access to high-quality foodstuffs containing only easily-digestible proteins. This should eliminate any carry-over of undigested protein to the caecum – thought to be a possible cause of early scours.

A ration must contain fats and carbohydrates as well as proteins. These are required to provide energy as well as maintenance. The carbohydrates are most important for producing energy, which includes body warmth; lack of them can cause stunted growth and loss of weight and condition. They are mainly provided by the grain portion of the horse's diet and are used to produce energy; any excess is stored as fat or excreted. To increase the utilisation of cereals and prevent any unnecessary wastage, (particularly in the case of older horses) the grain must be bruised, rolled or cooked.

Fibre, although of less importance to the horse than to the ruminant, is nevertheless necessary. Young horses and those doing fast work should, however, have rations in which the carbohydrate fraction is low in fibre. When stabled they should be fed good-quality first-crop hay, ideally cut as young as possible.

Minerals are necessary for all classes of horses and in most cases are supplied in adequate quantities by the food, nonetheless deficiencies can occur. An imbalance in the calcium/phosphorus ratio is the most common, in foals and young horses in general. This may produce round joints – see Chapter 3, page 48. Phosphorus should never be fed at a higher ratio than calcium. The minimum recommended ratio is 1.1 : 1.0, but this should be adjusted to 1.6 : 1.0 in cases where any calcium deficiency is suspected. A gross imbalance between these minerals can be determined if blood samples are taken by your veterinary surgeon for laboratory analysis. Some firms of proprietary horse feeds will have these blood samples tested for you free of charge if you use their food. It is a good idea to add calcium carbonate, calcium lactate or calcium gluconate to the ration of all young horses at the rate of about a tablespoonful (or 1 oz.) per day. Calcium lactate and gluconate are more readily absorbed than calcium carbonate, but far more expensive. Most plants contain large quantities of phosphorus, but not all of this is in a readily available form. Bran, for instance, contains high levels of phosphorus (1 :9.3 ratio).

The assimilation of calcium is governed by vitamin D (the sunlight vitamin), so any deficiency of this vitamin will produce a corresponding calcium imbalance in the affected animal. Therefore, during the winter months and in cases where a young horse has to be stabled for a long period for any reason, a vitamin D supplement might be necessary.

Vitamin A is naturally available from the carotene in grass and well-made hay; a deficiency of this vitamin can only occur when horses are kept without access to green herbage. Overfeeding vitamin A supplements can be harmful. Some people consider that vitamin E increases the fertility of breeding stock. It can easily be obtained in powder form for inclusion in feeds, but is very expensive. Vitamins can be administered by injection or by feeding, the latter being the less expensive method of the two, and the less likely to lead to over-dosing.

In conclusion: providing that both the quantity and quality of the food given is adequate it is very questionable whether there is any advantage to be gained by feeding additives.

The practical aspects of feeding are based on the theories already described, which can then be adapted to fall in with any particular stud's availability of produce and method of feeding.

However, all horses when stabled should be provided with salt licks so that they can help themselves whenever they wish.

In these days of high food prices some thought should be given to the advisability of bulk buying hay, straw and grain direct from farmers at the most favourable time, cost wise. This means buying in quantity possibly some six months or more before the product is used and having storage facilities for that period of time. However, it does guarantee a supply of fodder at a fixed price, and in recent winters has proved well worthwhile. Grain must be stored in rat-proof hoppers or buildings; if stored in sacks, plastic ones should not be used as these tend to make the grain mouldy, since there is no means of ventilation. Metal grain hoppers are probably the best investment. More than one hopper should be installed as each must be completely emptied before refilling, otherwise the most recently added grain will flow down through the middle, leaving old grain on the sides which will, over a period of time, become infested with weevil. A fan can be fitted to the hoppers, if desired, so that oats can be brought straight off the field and dried in situ.

The hoppers can be located over a grain crusher or the grain can be carried by means of a grain auger to the crusher. Oats should not be fed until the New Year after they have been harvested and are better still if kept for 12 months. Field beans should not be fed until they are at least a year old.

Only the very best-quality foodstuffs should be bought for young horses. This is of particular importance when it comes to hay, which should be well harvested, young, first crop, ryegrass/red clover mixture. Lucerne hay, if it has been well made, is also ideal for youngstock, as it has a high protein content and is high in calcium. Badly made lucerne or clover hay is unsuitable as it is very dusty and contains mould spores, which could cause broken-wind if fed to horses.

In cases where hay is made on the stud or brought straight off the field an in-barn hay drier should be installed so that, as far as possible, the quality of the hay can be guaranteed. Hay should not be used until after Christmas of the year of cutting at the earliest. If used sooner it will usually cause digestive upsets in horses of all ages.

Only good clean wheat straw should be used for bedding, ideally, as barley straw tends to make horses rub their manes and tails, which could spoil the appearance of a young horse sufficiently to affect his selling price or show-ring career for that season.

As an alternative to crushing or rolling, whole grain can be boiled before feeding. Whole barley can be boiled up with the linseed and whole field beans added for animals with a high protein requirement. These can be fed at the rate of:

½ lb. barley;
2 oz. field beans; } per horse per day.
1 oz. linseed

Mix with one part bran to six parts oats, the total quantity being adjusted to the individual horse. Soya bean meal may also be added at the rate of 5 oz. per day for weaned foals rising to 1 lb. per day (maximum) for yearlings or older horses with a high protein requirement. A little goes a long way, mixed into the ration.

Raw eggs can also be utilised in moderation for foals. They should be added to reconstituted milk powder and poured over the feed to dampen it. Dried milk powder should not be mixed with the feed and then dampened, as this will result in a paste-like mixture which is not very palatable. When feeding mashes, particularly those containing milk powder, great care must be taken to wash all the equipment and mangers every day. Any food which becomes stuck to the sides, rapidly goes sour and contaminates any fresh food with which it comes in contact.

When increasing the protein or bran level of the ration, always remember to increase the calcium level at the same time.

Flaked maize (which is high in starch) helps to put on flesh and keeps a ration open, thus increasing its palatability.

When feeding unweaned foals, care should be taken to ensure that the foal gets its fair share of the food. Greedy mares should be provided with an alternative manger and bad cases tied up while they are eating. An alternative method is to use a creep feeder as shown in Fig. 10 which can be attached to the manger.

An arrangement for the internal layout of stud loose-boxes is shown in Fig. 11. The feed section of the manger allows plenty of room for the food to be scattered the full length of the manger and for foals to get their share without too much free competition with their dams. The hay section on the right will accommodate $\frac{3}{4}$ bale of hay if necessary and is deep enough to prevent wastage. The hay is easily accessible to all sizes of foal, unlike hay racks and nets, which if high enough up the wall to be safe, are out of reach of all but

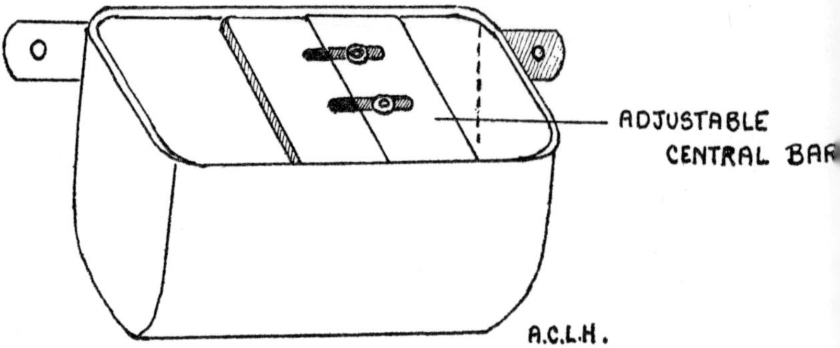

Fig. 10. Metal creep-feeder for unweaned foals enables a foal to eat without competition from its dam. Available from Horse Requisites, Newmarket, Suffolk

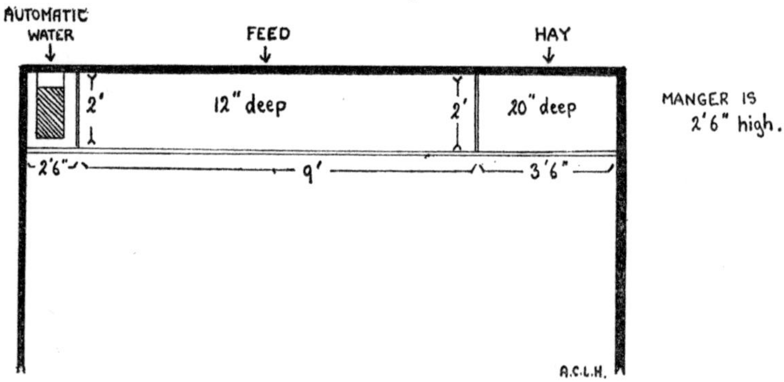

Fig. 11. An ideal layout for stud loose-boxes, suitable for all types of horses and ponies. Minimum floor area for ponies: 10 ft x 12 ft; for horses: 12 ft x 12 ft

the biggest foals.

When foals are at least two months old and the weather is warm enough they can be left out at night. This applies to Thoroughbred, Arabian and hunter type foals, as many native ponies live outside from the day they are born. Highly-bred foals and yearlings still need to be fed when they are running out, unless the quality of the grass is very good. Either they can be caught up every day and brought in for a feed or they can be fed out in the paddocks.

To prevent wastage the food must be kept off the ground and put into something which cannot be knocked over easily. A wooden field trough is an ideal arrangement; it should measure at least 10 feet long and be 2½ feet high. All exposed edges should be covered with metal strips to prevent damage by chewing. The troughs must be located several yards apart to prevent fighting. In the case of unweaned foals, it will be found that they will usually take up positions on the opposite side of the troughs to their mothers to avoid being chased away.

Many pony mares tend to suffer from laminitis, and this can present problems when an owner wishes to feed the foals and not the mares. In these cases, a creep feed can be built in the paddock, by constructing a square fence round some field troughs. The fence is made just high enough to allow the foals to walk underneath, but not high enough to let any mares through, so that the foals can receive a feed away from the mares.

Foals should consume 1 lb. of food for every month of age, so that by the time they are ready for weaning at about six months they should be consuming about 6 lb. of food a day, not including hay. Great care should be taken that no change is made in the foal's diet at weaning time. Before weaning, he should be introduced to some milk powder and a few raw eggs, if available. This will help to make good the loss of the mare's milk. Milk powder can be included in the foal's diet right up to the time the spring grass comes the following year. (EQUIVITE)

Part of the art of feeding young horses is to think of spring grass as the ideal food and then balance the ration according to the value of the grass available. When there is plenty of spring grass about no extra feed is necessary, but as the value of the grass declines through the year, so extra feed must be given to maintain an even rate of growth and development. Annual grass production for England and Ireland is shown in Fig. 12. It should be noted that grass

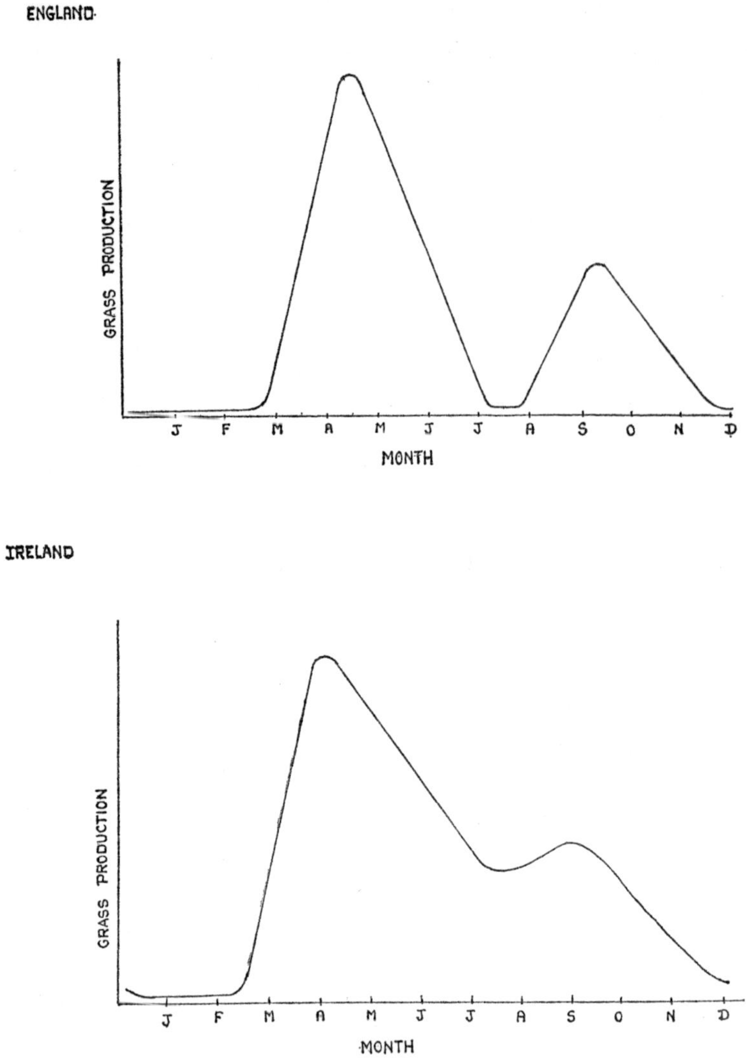

Fig. 12. Graphs showing average annual grass production for England and Ireland. The low trough between July and August on the U.K. graph is due to drought periods – more often seen in eastern areas of the country

which comes during the autumn flush is lower in protein and higher in fibre than spring grass.

At all costs, sudden dramatic changes in diet should be avoided otherwise the natural bowel flora will be upset and this can lead to digestive and metabolic disturbances, which in turn could cause conditions such as round joints. This occurs most commonly at weaning or after working heavily-infected stock.

Care must be taken to ensure that there is a constant supply of fresh water available for all horses. Water buckets, bowls and troughs should be cleaned out regularly and feeding troughs must be kept scrupulously clean at all times. To facilitate cleaning, corner mangers should be fitted so that they can be lifted out easily (see Fig. 13).

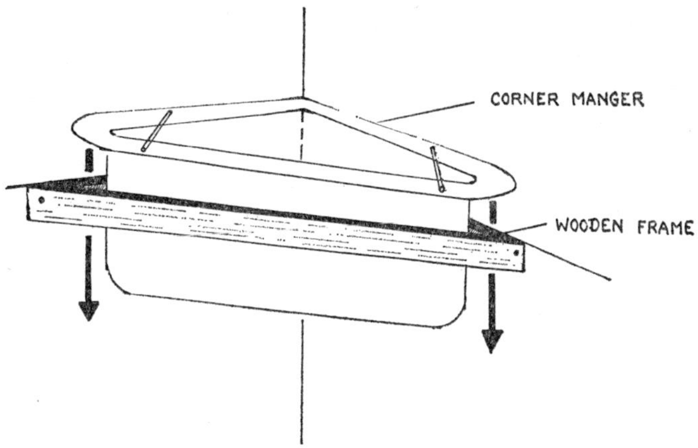

Fig. 13. Method of fitting a corner manger

7

THE MANAGEMENT OF WEANED FOALS AND YOUNG HORSES

Due to the cost of keeping horses these days, people are no longer prepared to give high prices for animals which are not absolutely sound, well grown and of good conformation. Also, as far as pedigree stock are concerned, they must be well bred. For anything less there is little or no demand. In the face of such a selective market the production of saleable young horses has become more of an exact science than ever before, and attention to detail the most important single factor.

It is well known that nobody manages stock quite as well as the owner himself, but failing the personal approach good staff is an absolutely essential factor. This means that a personal interest in the stock must be cultivated, usually by means of bonuses on sales and show-ring successes. Otherwise the production of yearlings as such is better not attempted these days.

When setting up a stud with the idea of producing yearlings for sale, thought should be given to the possible development of a 'yearling unit'. This can consist of a range of specially equipped loose-boxes, situated in a well-sheltered position, some distance from the main stud buildings and with easy access to well-fenced, sheltered, flat paddocks (see Chapter 1).

At the turn of the year, i.e. when the foals are about nine months old, the fillies should be separated from the colts. At this age the play of ungelded colts tends to get rougher, so purely from the point of view of safety they are better put into a paddock on their own and not left with the fillies. Except for native ponies, which will probably thrive as well if kept outside, all other youngstock should be stabled at night, at least from weaning until the warmer spring weather, when they too can be left out all the time.

For the last 2–3 months before the sales yearling colts should ideally be placed in separate, well-fenced paddocks, one animal to each paddock. Ungelded colts which are being shown should also be placed in separate paddocks to prevent injuries which can occur due to rough 'games' (see Plate 10). The paddocks used must be well fenced, but a word of warning here: the fencing must be high enough to prevent a colt from trying to jump out, but not so strong that he will seriously damage himself should he try. Care must also be taken when applying metal strips to paddock rails and gates; these should only be applied in short sections. I know of at least one fatal accident which was caused by a yearling attempting to jump out of his paddock and catching the top rail of the fence, which was unfortunately protected by a long metal strip. The rail broke, but the metal strip held firm, causing the yearling to turn a somersault and break his neck.

Foals grow fastest during the first month of their lives, but other periods of very fast growth occur between six and twelve months of age and after puberty. In the case of the normal spring-born foal this middle period of intense growth occurs between the months of October and April, when there is little, if any, grass available in the paddocks, particularly in the United Kingdom (Ireland has a somewhat longer growing season – see charts on page 82). Consequently if the young horses' maximum growth rate is to be maintained at this time, good-quality supplementary feeding must be given. This is of particular importance where stock are being reared for the show ring or sales.

Due to the current very high cost of all feed, there is a growing tendency among some breeders to attempt to cut production costs by concentrating on their better animals to the detriment of their less well-bred or well-made individuals. This is not to be recommended as a means of saving money, as under these conditions of management the reduced amount of food given is money wasted and the animal in question would show a better return if put down or sold as a foal, than produced a year later as a stunted yearling. Good individuals will always sell themselves, therefore more, not less, effort should be spent on the poorer yearlings, if they are kept at all, in order to attract a buyer at the sales.

Growth rate trials have shown that animals which have been kept on a lower plane of nutrition will eventually catch up with their better-fed companions, when put on to a correspondingly higher plane of nutrition, but not until, at the very least, a year later. And in the United Kingdom one cannot always rely on good spring or summer grazing to make good any winter feeding deficiencies. Also any sudden acceleration in growth rate can lead to the formation of round joints and should therefore be avoided.

When producing yearlings for the sales or show-ring, it does not pay even to attempt to cut corners. Attention to detail is of the utmost importance, as in order to attain their maximum potential, animals must be produced in the sales or show-ring in top-class condition, unblemished, with clean legs and joints and straight action. Therefore, although animals may be running out at grass both day and night, they should be caught-up daily and examined, even though they may not be brought inside for a feed. By doing this one has the opportunity to handle them regularly and so maintain their education. But above all it means that any accidental injury, however small, is seen and treated where necessary; also, a careful watch can be kept on the condition of the young horses' legs and feet. Any slight enlargement of the fetlock, knee or hock joints, should be especially looked for when the ground is hard. Should this happen, affected horses must be brought inside and put on to soft bedding, or turned into a straw or sand yard or a paddock which has been specially prepared and set aside with this contingency in mind (see Chapter 9, page 107).

By the time they are ready to be shown or sold, weaned foals and yearlings must be quiet to handle. Both the judges and buyers alike will want to examine them, and nothing tends to put a buyer off more quickly than an unmanageable or badly behaved young horse that he cannot get anywhere near. Having spent hundreds or even thousands of pounds on producing the horse, it would be a pity if a potential sale was lost for want of a little handling. This is particularly so when youngstock are bing reared on quiet private studs where they are constantly handled by the same few people every day, so some effort must be made to get them used to strangers. Any experienced person who visits the stud should be allowed to put a hand on each foal or yearling and possibly pick up their feet. It is debatable whether it is absolutely necessary to teach yearlings to tie up before they are sold. However if you do decide to tie yours up care must be taken to ensure that a breakable headcollar is used and that the rope is attached to a piece of baler twine and not straight on to a metal ring. This will ensure that something gives before the horse's neck, should he get frightened and run back. A young horse should never be left on his own when tied up in case an accident happens.

It usually helps if the young horse is given a small feed to eat while he is tied up for the first few times.

Youngstock must have their feet seen to regularly. It may not affect a horse's speed if he turns his feet in or out slightly, but it will affect his market value or show-ring career. Any gross deviation from the normal will put an extra, uneven strain on the legs and may also cause the horse to 'brush' or knock his legs and fetlock joints when moving, both of which could cause bony enlargements to develop. Either way it always pays to walk young horses out on a hard level surface, so that the blacksmith can check their action before he trims their feet, at regular monthly intervals. This way any deviation can be spotted and corrected immediately, before it becomes established. (See Chapter 3, page 55.) The actual shape of the feet is probably even more important and any tendency to contracted or 'boxy' feet should be especially looked for and if found must be worked on straight away.

Manes and tails should be kept in perfect condition. Foal's manes initially grow in an upward direction and nothing much can be done about this until they are at least four months old, at which age the hair is usually long enough to train over on to the off side of the neck. This is done by brushing it down with a damp body brush each day until the hair falls over naturally. Where foals are shown or sold before weaning, the mane should be left so that it does at least stand up slightly. This will help to improve the appearance of the top line of the neck, which is never very well developed at this age.

Young horses, particularly colts, which are running out at grass together often develop the bad habit of chewing one another's manes and tails. This can result in a partial 'crew cut' effect which will spoil the appearance of any horse. Therefore both manes and tails must be protected from the possibility of idle chewing. (Incidentally, this is also often indulged in by cattle running with horses.)

Possibly the most effective deterrent is a product called Renardine, which can be bought from hardware stores. This is a very nasty-smelling liquid which should be mixed with water or oil. Use an old bucket which you can keep especially for this purpose and wear a pair of rubber gloves. Apply the mixture to the horse's manes and tails about twice a week and you can guarantee that they will not attempt to chew one another again.

This product does not prevent young horses from rubbing their manes and tails. If they start to do this excessively, the exact cause must be established before any treatment can be given. Some horses, particularly ponies, are allergic to the effects of certain flies which are prevalent during the summer months. This is known as 'Sweet Itch', and affected animals must be brought inside during the day-time and placed in stables which are equipped with an effective fly-killing device.

Tail rubbing alone can be due to the effect of worms, in particular the seat worm (*Oxyuris equi*). The female lays her eggs around the anus, which sets up an irritation causing the horse to rub these parts. (See Chapter 8, page 94.) Another possible cause could be the presence of mites in the mane or tail, in which case an application of benzyl benzoate or tetmosol, applied undiluted daily to the affected area, will usually cure the condition.

8

THE PROBLEM OF PARASITES

INTERNAL PARASITES

Worms are parasites. A parasite is an organism which lives on or in another organism (in this case the horse) deriving its nourishment from the host animal without actually killing it. If the parasite does destroy its host it is defeating its own object by depriving itself of a place to live and reproduce its species. Therefore where an animal dies as a direct result of a worm infestation it is a grave error on the part of the worms and not the inevitable result in a natural course of events.

Why then should we worry too much about worm parasites? The reason is that horses which have not previously come in contact with worm infections are not immune to their effects – while other immune but infected horses are capable of infecting all other horses with which they are turned out. The external signs of parasitism are:

(i) Loss of weight
(ii) Loss of condition, poor coat
(iii) Debility
(iv) Anaemia
(v) Colic
(vi) Death

At all costs one should avoid letting paddocks become infested with worm parasites. Although immune adult horses

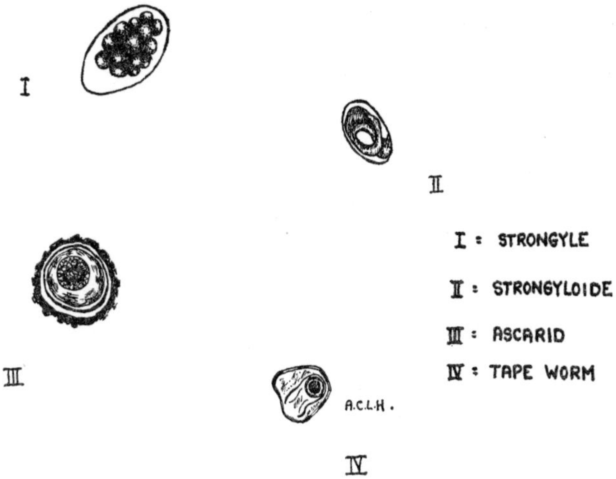

Fig. 14. Individual worm eggs as they appear under a microscope

will probably show little if any signs of infection, the young-stock, despite good feeding will always look poor and stunted, with the result that economically they are completely unviable. They will consume as much, if not more food than uninfected animals, but have a very low food conversion ratio, which will result in poor condition and stunted growth but with normal costs of production. After worming, such stock may develop round joints.

The worm parasites usually involved are:

(i) Red worms, large and small
(ii) Round worms
(iii) Strongyloides
(iv) Tapeworms
(v) Seat worms
(vi) Lungworms
(vii) Liver fluke
(viii) Bots.

A complete list of these worms and their various degrees of pathogenicity is given on the following page:

Common Internal Parasites of the Horse

Group		Common Name	Proper Name	Degree of Pathogenicity
1	(a)	Large red worms	*Strongylus vulgaris*	++++
			Strongylus equinus	++
			Strongylus edentatus	++
	(b)	Small red worms	*Trichonema*	+
			Triodontophorus	+
			Trichostrongylus axei (common also to cattle and sheep)	+
	(c)	Round worm	*Parascaris equorum*	+++
	(d)	Strongyloides	*Strongyloides westeri*	+
	(e)	Seat worm	*Oxyuris equi*	±
2		Tapeworm	*Anoplocephala pertoliata*	±
3		Lungworm	*Dictyocaulus arnfieldi*	++
4		Liver fluke	*Fasciola hepatica*	±
5		Bots	*Gastrophilus*	±

The strongyles are by far the most common and important species and are in the family of the red worm. The adults live in the intestines, where the females lay their eggs, which pass out in the droppings on to the grass. They hatch out into larvae and climb up the blades of grass, where they hope to be eaten by their particular host animal. Any larvae which are fortunate enough to be eaten pass with the grass into the animal's stomach. If the animal is their particular host species they will survive; if not, they are destroyed. Larvae which survive migrate through the intestinal walls, travelling up the mesenteric arteries; that is, except for the small strongyle which remains in the walls of the intestines. The mature larvae eventually return to the intestines, become adults and the females commence egg laying. (See Fig. 15.)

The complete cycle takes about six months. Migrating larvae can cause aneurisms to form at the root of the

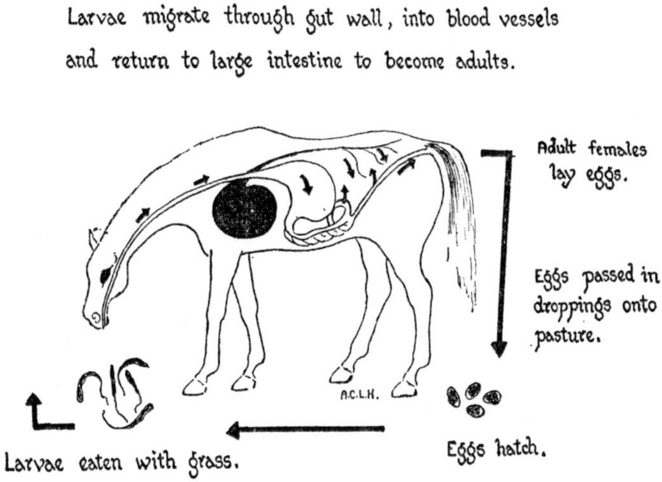

Fig. 15. The life cycle of the red worm

mesenteric arteries, which may result in the sudden death of the horse, or cause damage to part of the gut, which results in chronic intermittent colic.

Another type of worm most commonly associated with young horses and foals, is the ascarid (round worm). These can easily be seen in the droppings of infected stock as they measure up to 14 in. long. Where very large infections of this worm occur they can cause intestinal obstructions and perforation of the bowel, the latter resulting in a fatal peritonitis. This worm is most common in horses under three years old; older horses appear to have a high resistance to ascarid infection. The eggs are passed in the droppings on to the pasture in the same way as strongyle eggs but in this case they do not hatch but are eaten with the grass and develop inside the host animal – the life cycle of the round worm can be seen in Fig. 16. The complete life cycle takes about 60 days.

Strongyloides are common in young foals. They appear to be self-limiting, as they are seldom found in the droppings of older animals; they are therefore of little consequence,

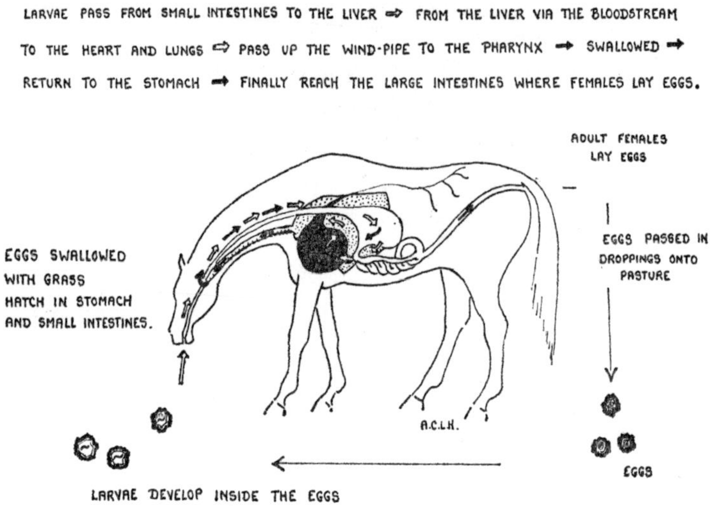

Fig. 16. The life cycle of the round worm

although some authorities maintain that they can be responsible for scouring in foals.

The eggs of the seat worm (threadworm) are seldom, if ever, found in faeces samples sent for laboratory analysis, due to the fact that the female worm lays its eggs around the anus and not in the intestine like the previously-described species of worm parasites. The eggs of the seat worm usually set up an irritation causing the horse to rub its tail and hind-quarters.

Tapeworms seldom cause much trouble in horses unless there is a large infestation.

Lungworms are most often found in the donkey, which appears to be their natural host, and seldom produce symptoms unless accompanied by a secondary infection. In the horse, however, they often produce symptoms of coughing and broken wind – even in the case of small infections. Where lungworm occurs in the horse it usually originates from an infected donkey. (See Fig. 17.)

The life cycle of the lungworm differs from those already

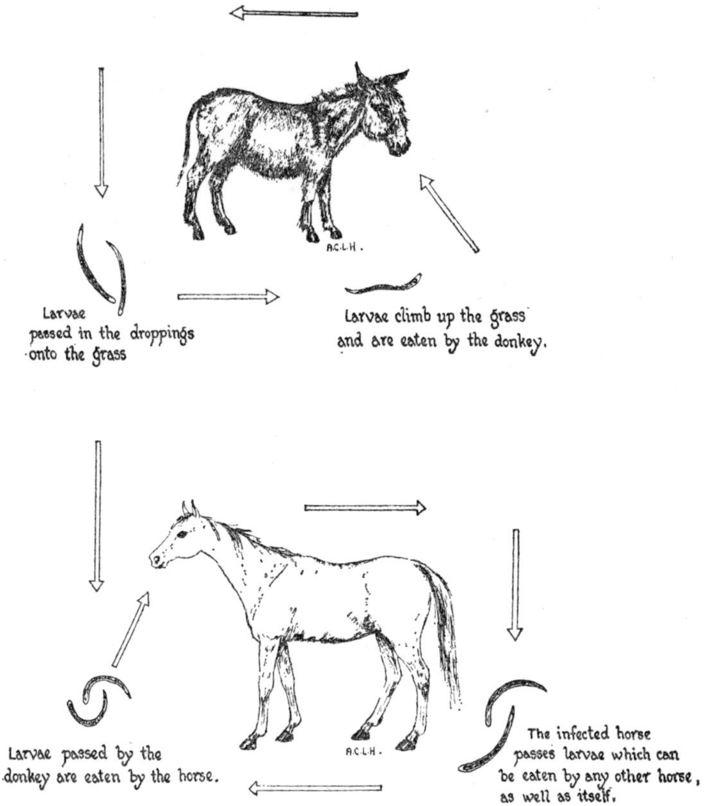

Fig. 17. The life cycle of the lungworm

mentioned in that although the female lays her eggs in the intestines, they hatch by the time they are passed in the droppings. On the pasture the larvae mature and are eaten with the grass. When eaten by the horse or donkey, they pass down into the intestines. From the intestines they enter the lymphatic system and so reach the right side of the heart. From the right ventricle they become blood-borne and migrate to the lungs; here they break out into the air passages and eventually reach the windpipe, when they are either coughed up and swallowed or pass up in the mucous and are swallowed, returning to the intestines were the females commence egg laying. The worm found in the horse is *Dictyocaulus arnfieldi* which, in common with most other equine worms, is specific and differs from those causing lungworm infections in other classes of farm stock.

Horses are sometimes infected by liver fluke, but only in areas where these parasites are found in other stock because the intermediary snail host is present.

Bots are normally not so much a problem in the British Isles and Ireland as they are in some other parts of the world. The bot fly attacks the horse by laying eggs on its forelegs, where the horse can lick them off easily. The eggs hatch into larvae in the horse's stomach and attach themselves to the walls where they remain for about 12 months. Large numbers of larvae can cause blockages, impaired digestion, colic and sometimes rupture of the stomach. To control infection, eggs should be removed from the horse's coat and larvae eliminated from the stomach by treatment with carbon disulphide or Duex organo-phosphorus compound. Ordinary anthelmintics do not usually have any effect on bot larvae.

EXTERNAL PARASITES

Lice are often associated with worm infections in horses. They are more prevalent during warm wet winters than they are when the weather is really cold and frosty. Two distinct varieties are found on the horse: the biting louse and the sucking louse. Lice cause considerable irritation and affected horses will soon rub themselves raw in patches and lose condition rapidly. If the horse's coat is turned back, particularly in the areas of the mane and tail, lice can usually be seen with the naked eye.

Liberal dressings with louse powder, repeated at weekly intervals until all trace of lice have disappeared, are usually all that are required. Care should be taken not to get any louse powder in the animal's eyes as it will cause inflammation.

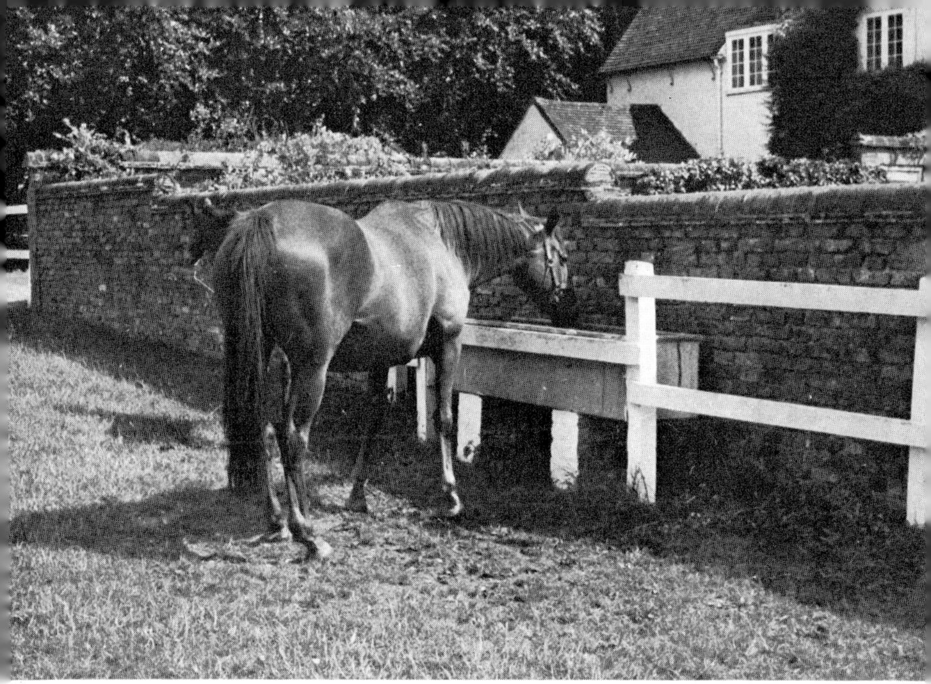

1. To save labour, mains water should be laid on to all the paddocks. Troughs should be sited away from gateways, in a sheltered position.

2. Topping the grass regularly throughout the growing season is essential when mixed grazing is not practised.

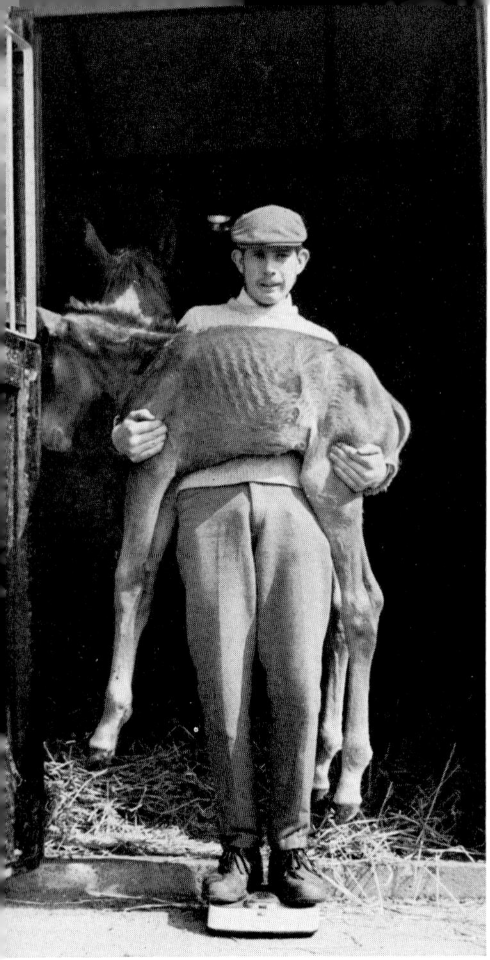

3. The correct method of lifting and weighing a new-born foal. Large Thoroughbred foals will weigh as much as 120 lbs or more; very small Thoroughbred foals less than 90 lb.

4. The Eastrac Paddock-Vac. With the help of this machine, droppings can be removed from the paddocks more quickly and more efficiently than by the traditional method, particularly if the grass is kept short.

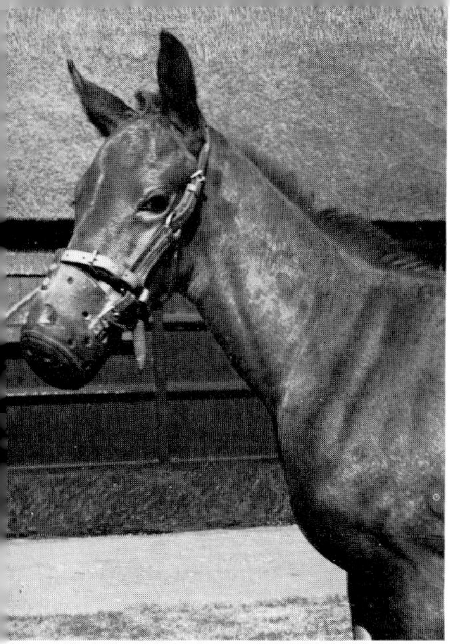

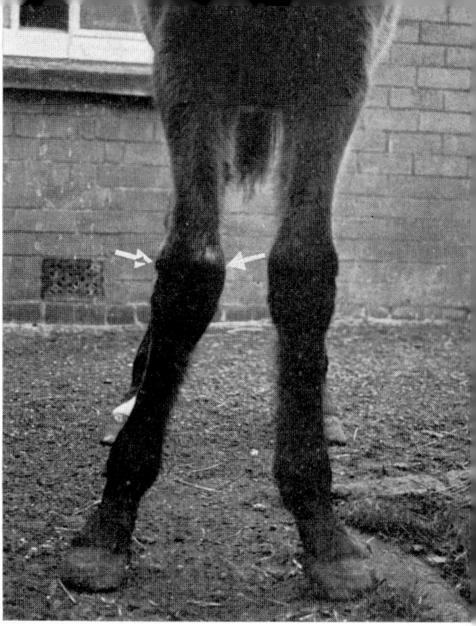

5. (*above left*) If haemolytic disease is to be prevented, a muzzle must be put on affected foals the moment they get to their feet.

6. (*above right*) Uneven growth of the epiphysis. The near foreleg is bent outwards from the point indicated by the arrows. The arrows also show the location of the epiphysis at the end of the radius and the swellings at this point which are due to epiphysitis.

7 & 8. Characteristic abnormal lying positions of a foal with meconium colic.

9. Paddock weaning – the last mare can be left with the weaned foals for several weeks.

10. Two yearling colts at play in their paddock.

11. Yearlings being led round prior to being sold at Tattersalls Park Paddocks, Newmarket. Note the neatly turned out horses and grooms in the foreground. (*Photograph by Peter Faulkes.*)

12. A horse ready to travel by road. When horses travel by air a poll guard should also be used.

13. The current fashion is to make a large number of very small button plaits. Also note the neatly plaited tail. A first-prize winning, two-year-old filly at the Dublin Horse Show.

14. The judge is now sorting out his class and has just started to pull the horses he likes best into the middle of the ring.

15. Wait until the ring steward tells you to bring your horse out for the judge to see, then stand him sideways on to the crowd and in front of the judge.

16. Walk to the far end of the ring, turn and trot back. Once the horse is trotting well, drop your hand so that the horse is going on a slack rein. In this way he will have to carry himself and not lean on the bit.

17. Native ponies are shown unplaited.

18. What to aim for – a top prize winner at the Dublin Horse Show.

Ringworm is a highly contagious fungal skin disease of horses and cattle which is transmittable to humans. Care should therefore be taken when handling animals with this disease.

Affected horses will first be seen to have small slightly-raised round spots from which, in a matter of a few days, the hair drops out, leaving small (round) bare patches – hence the name. These will increase in diameter if left untreated and become crusty.

If you suspect that a horse has ringworm, it should be isolated immediately and your veterinary surgeon consulted as to the best treatment.

Horses, like cattle, can be attacked by warble flies, which lay their eggs on the horse's legs. The larvae hatch and burrow under the skin. As they mature they make their way towards the animal's back, where they eventually form a small lump with a characteristic small hole in the middle through which the larvae breaths and through which it eventually hatches. Larvae should be removed surgically by your veterinary surgeon, before they have time to hatch.

PREVENTING INTERNAL PARASITES

Methods of preventing an internal parasitic problem on studs are :

(1) *Regular dosing of all animals*

All foals should be wormed against both strongyle and ascarid infections from the time they are two months old at regular monthly intervals. It is recommended that young foals are tubed by a veterinary surgeon as this ensures they receive the correct dose. Any older horses which will not clean up a feed containing a worm dose should also be tubed or have the dose squirted on to their tongues using a syringe. Care should be taken to check that animals have eaten up their food completely whenever they are wormed, so that you know they have had their full dose of worm medicine.

Every horse on the stud must be included in the worming programme for it to be successful, even if it is only a retired hunter or pony. Animals running at grass must be caught up and wormed regularly every eight weeks.

Worm parasites have the ability, in order to save themselves, of becoming immune to the effects of one particular drug if this is used continuously for a long period of time. Therefore to be really effective the drug used must be changed from time to time. This will have the effect of not allowing the worms sufficient time to develop any immunity and so an effective programme of worm destruction is maintained.

Most authorities, at the moment, recommend the alternate use of Equizole or Telmin and Strongid. Equizole and Telmin are virtually the same drug. These drugs control strongyles in both young and adult horses. It is only necessary to dose adult stock against strongyles, unless you have reason to believe that an individual is also infected by another species of worm. However, young horses are usually infected by ascarids and must be treated for this type of worm as well as for strongyles in which case Strongid or Piperazine can be used.

In this way all the horses will be kept free from egg-laying adult worms in their intestines, and your pastures will remain relatively clean.

(2) *Paddock management*

Ideally all droppings should be removed from the paddocks daily, but with the present high cost of labour this is not always a practical proposition. However, every effort should be made to remove droppings from the smaller or most heavily-used paddocks, otherwise they will tend to become 'horse-sick'. (See Plate 4.)

From the point of view of worm control, paddocks should be rested for the six-month winter period, when frost will drastically reduce the numbers of surviving larvae.

Horses are very selective in their grazing habits and will

only graze areas which have not been contaminated by their own droppings, which is a limited natural safeguard against self infection with worm parasites. Large areas of paddock will tend to be rejected by the horses with the result that unless the grass is kept topped regularly it may grow very long and coarse in places. As an alternative to topping, a herd of de-horned or naturally polled cattle may be grazed with the horses. They are less selective in their grazing habits and will clear the areas of long grass left by the horses. Apart from *Trichostrongylus axei,* which is parasitic for cattle, sheep and horses, any worm larvae which are consumed by the cattle will be destroyed and so the number of worm larvae on the paddocks will be greatly reduced. Ideally one should maintain a grazing rate of one-fifth horses to cattle, but this is not always feasible.

The practice of harrowing paddocks to scatter the droppings only has the effect of contaminating larger areas and reducing the clean area available to the horses. Ideally droppings should be removed from the paddocks at least once a week and the paddocks rested for a minimum of three consecutive months each year.

(3) *Routine analysis of droppings samples**

Although horses may be wormed at regular intervals, it does not necessarily mean that they are no longer passing any worm eggs in their droppings. Some worms manage to develop an immunity to a particular drug or drugs and will continue to survive despite treatment. Therefore a periodic check should be made to make sure that all animals are free from worms.

For faecal worm egg counts the samples must be as fresh as possible. They should be placed in an air-tight container, such as an Equizole tin or sealed polythene bag. They must

*For details of how to carry out a simple faecal egg count test see Chapter 16 of *Stallion Management,* A. C. Leighton Hardman, Pelham Books.

not be allowed to dry out and must be kept as cool as possible so that the eggs will not hatch before they can be counted. The sooner the samples are taken or sent to the laboratory for testing the better. Samples should be taken from the centre of droppings and should represent more than one pile of droppings from the same horse, if possible. They must be clearly marked with the horse's identity at the time of collection.

For a straightforward worm egg count, an Equizole tin full of droppings is sufficient but if a lungworm larvae count is required, at least the quantity that will absolutely fill a $\frac{1}{2}$ lb. honey jar is required.

9

THE PRODUCTION OF FOALS AND YEARLINGS FOR SALE

First of all decide which sale you are going to send your animals to and make sure to book the number of places required in good time. The major yearling and foal sales in England and Ireland are as follows:

Yearling sales
Tattersalls Houghton Sales, Newmarket
Goffs Premier Yearling Sales, Kill
Tattersalls October Sales, Newmarket
St Leger Yearling Sales, Doncaster
Goffs October Sales, Kill
September Yearling Sales, Ballsbridge

Foal sales
Tattersalls December Sales, Newmarket
Goffs Second November Breeding Stock Sales, Kill
Great Annual November Sales, Doncaster
December Sales, Ballsbridge

Not all sales are open to new vendors, so this point should be checked before any plans are made. Having ascertained which sales are open to you, the decision as to which sale or sales to patronise must be made. Established breeders usually send their stock to the same major sales each year, so the buyers know, from one year to the next, exactly which sale and on which day their stock will be sold.

Read the conditions of sale carefully before you make your entries. These are supplied by the auctioneers on request and are usually sent with the entry forms. Basically they are the same for all the firms, but often differ in some minor details. When applicable, particular note should be made of the information which must be supplied by the vendor relating to horses which are wind-suckers, crib-biters and weavers, or have been operated on for any unsoundness of wind. The pedigree details which appear in the catalogue can influence the price an animal makes, so it really pays to have your pedigrees compiled by a professional who has access to all the international racing results.

In order to make a name for your stud and be a success you should only submit your best animals to fill your places in a major sale. Inferior individuals which would spoil the overall standard of the draft should be submitted elsewhere. If one year you only have poor individuals to offer, safeguard your stud's good name and send them to a minor sale. Contrary to popular belief, by doing this you will not lose your place in the major sales. It is sometimes difficult to take an objective view of your own stock, therefore in case of doubt it is advisable to ask the auctioneers or a bloodstock agent for an opinion of your horses and abide by the decision.

Both the individual sale and the time of day the lots are sold has a bearing on the price the animals make. A really high-class individual will probably sell well, in the right sale, at whatever time of day it comes up, but average individuals are affected by the trade, which tends to vary throughout the sales. The highest prices are usually obtained during the afternoon and evening sessions of the middle days of the sales. Most sale companies allot these places to clients with the longest standing. However, some advantage can usually be gained by negotiating with the auctioneers to ensure that your draft comes up at a time when other good animals are going through the ring, which at least ensures the continuing interest of the buyers. Goffs have devised an alphabetical place system for their sales at Kildare Paddocks, Kill, Co Kildare. In 1975 they picked a starting letter for each sale, which is moved forward five letters each year. The vendors are catalogued under the first letter of their own or their stud's name and so everyone gets the same chance to sell at the best times over a period of years.

Part of the art of producing foals and yearlings for sale is to put them in the right sale at the right moment. If, for instance, you have bred a foal which is very closely related to a high-class winner that year, it would be better to offer it for sale as a foal, than to wait until the following year, when people may have forgotten about the winning relation.

Overseas buyers attend most of the major sales of both yearlings and foals, whereas the smaller sales appeal to the home market only. Individuals should therefore be carefully selected with particular sales in mind. An animal sired by a stallion who was a great success in the north of England in his racing days, will have more appeal to northern trainers, and will usually sell better at Doncaster than it would in the south of England. A good individual with a high-class pedigree of international appeal should be aimed at a major sale where overseas buyers are in attendance.

Foals are usually sold after weaning, during the months of November and December, by which time they will have their full winter coats and nothing either can or should be done about this. There is little point in entering foals born later than the middle of May for the autumn sales that year, as those born after this time will generally be too small and immature to attract buyers. Foals which are going to be sold should be weaned by the first week in September, at the latest, to enable them to recover from being weaned and leave enough time to prepare them for the sales.

Sale foals must be brought in every night after weaning

but can be left out all day. So long as they are in good condition, it does not matter if they have a full winter coat, providing it is shiny.

Animals which are going to be shown or sold should never be bedded down with barley straw, as this tends to make them rub themselves. Wheat or rye straw makes the best bedding material; failing this peat moss or shavings could be used, but buyers tend to travel round the studs examining the lots some time before the sale and nothing shows a horse off better than a deep bed of straw.

As soon as the foals have settled down after weaning, their manes should be trained over on to the off side of their necks and about four weeks before the sale they should have their manes pulled. At this age the mane is seldom very thick, so if preferred, it can be pulled by using your finger and thumb or a mane comb and a sharp knife. Brush the mane out so that it lies in a level line down the neck. Pick up a small section of the mane at a time, as you would for pulling in the normal way, (see page 146), back-comb, and instead of pulling the long hairs out altogether, simply break them off against the blade of the knife. Continue down the neck. In most cases it is better to leave the forelock unpulled.

If you have any equipment to buy do not leave it to the very last minute – you might not be able to find exactly what you want at the time and that could be a disaster! Start collecting everything together about 2–3 months before the sale – you will be surprised just how quickly time can fly.

Most foals and yearlings look better and are more controllable if they are presented at the sales bitted and wearing a neat brass-mounted in-hand bridle. These bridles should be double-stitched throughout, and look neatest if made up with narrow ($\frac{1}{2}$ in.) cheek-pieces. For the best effect, the leather should be cleaned with dark tan shoe polish and finished with a soft cloth to a brilliant shine. The brass buckles and bit rings must also be polished to a maximum shine. A leather leading rein with a brass buckle both looks better and will last longer than a white web rein. As with the bridle, clean the outside of the rein with dark tan shoe polish; only saddle soap the reverse side. Reins can be made with Y couplings, which are attached to either side of the bit, or separate bit couplings used with an ordinary rein.

A short leather strap attached to the centre ring at the back of the coupling and to the noseband of the bridle will help to transfer some of the pull from the horse's mouth to his nose. Foals and some yearlings are better lead with a flexible rubber bit. Although some people use small nylon bits, I personally find the rubber bit less likely to damage the young horse's mouth if he plays up.

Once the foals are weaned their food should be gradually increased until like the yearlings they are receiving four feeds per day: you will be surprised just how much food a foal is capable of eating, particularly by the end of November, and you may be even more surprised by the amount of food a yearling can put away.

The way an individual foal or yearling looks at the sales, depends directly on the way he has been fed. Plenty of flesh on the back, neck and quarters, will make most animals look better and tends to cover up some faults in conformation. The other factor which influences the looks of an animal to a large extent is the shine on his coat. This is mainly achieved by including some oily products in the ration. Those most commonly used are:

Linseed – normally fed boiled, but pure linseed oil (for animal feeding) can be used instead. When preparing animals for sale, it is a good idea to feed some linseed every night. It takes about a month before the effects of any oil in the diet can be seen in a horse's coat, so this ration must be fed for at least 6–8 weeks before a sale, to have the maximum effect.

Cod Liver Oil – some horses do not like the taste of pure cod liver oil, but most will accept cod liver oil and malt very readily. This product contains vitamins A and D, and as such is highly valuable for stock which are being kept inside and which might otherwise be short of these vitamins. A large tablespoonful per day is usually sufficient.

Olive Oil – this is practically a pure fat; it doesn't have a very distinctive flavour, so is usually accepted by most horses.

Seaweed Meal – although by no means an oily substance, this does help to put a shine on a horse's coat. It contains high levels of iodine and can be obtained from most saddlers.

Having achieved the all-important shine from the inside, by means of feeding, there is little problem in getting a lovely shine externally on the horse's coat any time you wish. That is without the necessity of having to spend hours on full-scale grooming or strapping of the horse, a great saving in labour, which is always to be recommended.

Yearlings are normally sold during the months of September and October when, if left to nature, they are just growing their winter coats. This tends to make the hair look coarse and dull, even with the aid of a special diet. Therefore one of the principal aims, when producing yearlings for sale, should be to maintain their full summer coats right up to the day of the sale. This is usually not too difficult to achieve for the early September sales, but becomes quite an art where the mid- or late-October sales are concerned.

Depending on local climatic conditions horses in the British Isles usually start to grow their winter coats any time from the end of July onwards. Sale yearlings should therefore be brought inside before the end of July if their full summer coats are to be kept.

The loose-boxes used for sale yearlings must be as warm as possible, (see Chapter 1). This will help to preserve their coats, but adequate ventilation must be provided at all times, otherwise there is a very big danger that the yearlings will develop snotty noses. It is better to use night rugs than to shut the loose-box top doors. Some people use infra-red lamps, but as these will not be available at the sale there is every chance that the yearling will feel cold, especially as the loose-box at the sales may well not be as warm as his loose-box at home, even without the lamp. If a horse feels cold he will look miserable and his coat will stand on end and appear dull, which must be avoided at all costs.

Once the yearlings have been brought inside, they must be walked out every day; initially this need only amount to half an hour. Weather permitting, they can be allowed out into a paddock, preferably singly, for an hour or two each per day. If the ground is hard and you wish to give them more than just walking exercise they can be turned into a straw or sand yard or better still a paddock which has been disced and harrowed down to a fine seed bed. This will provide a 'play ground' for the yearlings with good going even when the paddocks are rock hard. The main idea is to get them used to being led, and reasonably fit without going to the extent of walking any fat off them. The same applies to foals, but due to their extreme youth, they should never be walked for longer than about half an hour at a time. The amount of exercise each individual animal receives must be carefully regulated against the amount and quality of food he is receiving, so that he is well covered and in reasonably hard condition by the time he reaches the sale. It is unnecessary to walk yearlings for more than one hour each per day to achieve this optimum state. One man cannot reasonably be expected to prepare more than three yearlings at a time for the sales.

The sooner animals are mouthed when they are being walked the better. There is then less likelihood that any of the yearlings will manage to get free and perhaps damage themselves at the eleventh hour.

In order to mouth a foal or yearling, you will need a straight bar mouthing bit with keys (see Fig. 18). This can be

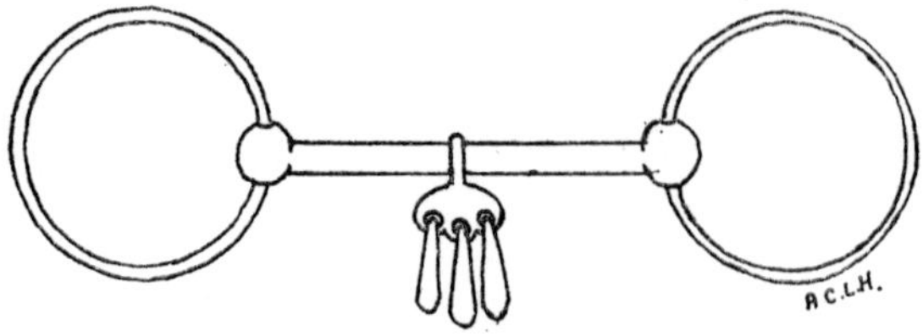

Fig. 18. Mouthing bit

attached to the existing headcollar with bit straps or clips, or be tied in place with baler twine. The fewer people involved in putting a bit into the horse's mouth for the first few times, the easier it usually is. In other words, if one person can manage to do the job quietly, on his own, this is better than having two or three people to do the same job – which can be psychologically bad for the horse.

To put a bit in a young horse's mouth you should attach the bit to the off side of the headcollar; keep hold of the bit all the time, so that it does not hit the horse on the side of the face should he throw his head up in the air for any reason. Move round to the near side, hold the bit in the flat of your left hand and gently insert your thumb into the corner of his mouth on this side, which will cause him to relax his jaw. Then you can slip the bit into his mouth, without catching his teeth, and secure it to the headcollar on the near side. Never try to force the bit between the horse's teeth.

If two people put the bit in the young horse's mouth they should take up positions one on each side of the horse's head and hold the ring of the bit and headcollar on the side nearest to them. Then they should gently open the horse's mouth as described above, slide the bit into his mouth simultaneously, and attach it to the headcollar on either side.

For young horses it is preferable to adjust the bit rather too high in their mouths, so that it just wrinkles the corners of the mouth. This prevents any chance of their learning the bad habit of getting their tongue over the bit, especially during the stage of mouthing when they are playing with the bit and keys.

Mouthing should be carried out in a loose-box, but first check that there are no nails or projections on which the rings of the bit could get caught. If there is no cage over the door the top door should be shut so that the horse cannot get himself hooked up on the door fastenings, and possibly break his neck.

The bit should only be left in the mouth for about half an hour on the first day, but it can be left in place for increasing periods of up to an hour over the following few days. It usually takes about two weeks to mouth a horse, after which time he can be led out from the bit. At this point I usually substitute a flexible rubber bit for the metal mouthing bit.

As a general rule youngstock should not be led with a short leading rein, as this does not give the person sufficient chance to hold on to the horse should he really play up. Reins measuring at least 12 ft. or more in length are safest. Where more than one animal is being led out at a time at least one horse's length should be kept between each so that, in the event of one kicking out suddenly, he will not make contact with the horse immediately behind or with its leader.

Initially, walking should take place in a quiet paddock, away from other horses, with the *gate shut*, so that should an animal get free, it will not be able to gallop out on to a road or round the stud. Until they are shod, yearlings should not be walked for any length of time on a hard surface, otherwise they will become foot sore. For this reason foals, which are never sent to the sales shod, should only be walked on grass or other soft going. It is customary to shoe yearlings in front only. Apart from the protection it affords, a shod foot always looks neater than one which is left unshod. Most sale rings have hard going, so when your blacksmith comes to shoe the yearlings before the sale he should be asked to leave their hind feet slightly on the long side, so that they do not go foot-sore behind while being walked round. Animals which still toe in or out, at the time of the sale, can have their looks improved slightly by off-setting the centre clip on the affected foot, and also, of course by leaving the foot slightly longer on one side than the other to help straighten the toe.

For the last month before the sales yearlings are probably better kept inside and just walked out each day for exercise, so there is less chance of their damaging themselves at the last minute. If they start to get too fresh for the men to handle easily, they can be lunged, but a word of warning here – most young horses are not very well balanced on the lunge and usually tend to fool around. So if you do decide to lunge your yearlings their legs must be very well protected with bandages and boots if knocks, bony enlargements and lameness are to be avoided.

In order to attract the buyers at the sale, all the horses must be taught to walk out smartly, with as long a stride as possible. A horse which walks slowly never quite fires the imagination to the same extent as one which goes round the sale ring at a good, easy striding, active walk. In order to achieve this, the horse must be schooled to walk on when asked and particularly to walk well on his own. This can be done in several ways, but possibly the easiest is to allow your people to carry a long stick (about the length of a dressage whip) in their left hands when they are walking the young horses initially. When a horse hangs back or fails to walk on, the handler should click or give a word of command to 'walk on'. If this is ignored by the horse, he should repeat the word and at the same time give the horse a sharp reminder on his quarters with the stick, without altering his position by the horse's shoulder; hence the length of stick. Horses soon learn to associate the one with the other and will very quickly come to obey the voice alone, at which stage the stick can be left at home. While they are being walked round, the young horses should be pulled out in turn, made to stand correctly and walk away from their companions since they may well be required to do this at the sale.

Sometime during the last month both yearlings and foals should have their manes pulled or trimmed as described in Chapter 14, page 146. Foals' tails must be left in their natural state but yearlings should have their tails pulled; this will help accentuate the roundness of their quarters. Tail pulling is better done over a period of time so that you do not make the top of the tail sore. A tail bandage should be put on during the day to keep the hairs at the top of the tail looking neat and tidy. The bottom of the tail should be cut level so that it comes just below the hocks when the horse is moving. To achieve this, use a pair of very sharp scissors and cut the end of the tail on a level with the bottom of the hock joint and at an angle sloping upwards slightly from the back to the front – when the tail is hanging naturally it will then appear to be level at the bottom. With some yearlings it is only necessary to cut about 1 in. from the end of their tails. Any long hairs which may have grown in the ears should also be removed.

When horses are inspected by the buyers at the sale they will expect your staff to be able to stand the horses up for them correctly (see cover photograph) and walk or trot them out as shown in Fig. 19. When trotting a horse out for a buyer, as a general rule it is better not to ask the horse to go too fast, as some horses tend to move wide behind if asked to trot very fast. When trotting always remember to slow down to a walk before you turn to trot back. *Always walk round* the horse and never ask the horse to walk round you; by doing this you will help to keep him balanced and not cause him to move crookedly. For the same reason you should walk about three or four strides before asking the horse to trot on again.

Running a horse up in-hand must be practised at home before the sale and if necessary a stick can be carried to teach the horse to 'trot on' when required.

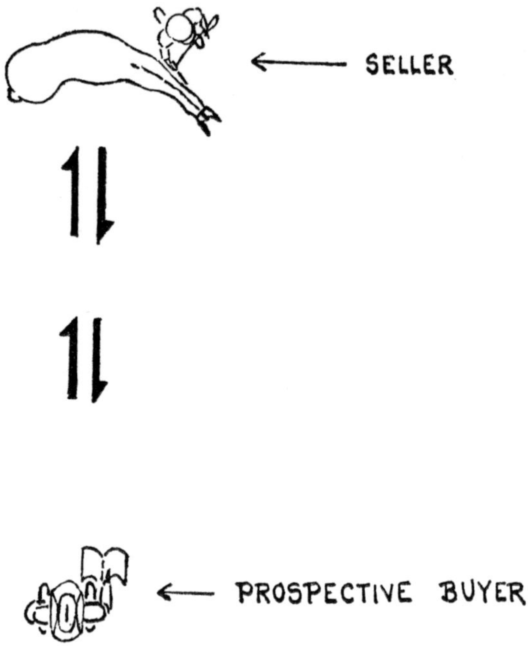

Fig. 19. Correct way to walk a horse out for inspection. When you turn, always walk round the horse – never ask the horse to walk round you

Some weeks before the sale you will be approached by the various horse magazines and newspapers to advertise your stock. They will usually compile the advertisement for you but make sure they keep it simple and easy to read. It can pay to advertise in the leading magazines and papers but do not overdo it, as wholesale advertising can be expensive.

At most sales it is customary, and also a great help to the buyers, if you have cards printed and attached to the loose-box doors giving details of the individual lots. These usually measure about 12 in. by 20 in. and again should be kept as simple as possible.

Having made quite certain that your stock will look as well as they possibly can on the day, you should complete the picture by making sure that your staff will be equally well turned out. When you are offering a draft of animals, they will create a better impression if you make sure that everyone is dressed alike. Buy new smocks and only allow your men to wear wellingtons if the weather is very wet. Make sure their shoes are clean and polished.

Make arrangements for the animals to arrive at the sale as early as possible. For small sales they should arrive the day before, but for larger sales at least three days before they are sold, to allow them sufficient time to settle down in their new surroundings. Where animals are being brought over to a sale by ferry, at least one week should be allowed for bad weather hold-ups. Check that you have the correct Ministry documents well before you are due to sail. You must also check what, if anything, the various sales companies provide for their vendors. All the major bloodstock sales companies provide bedding and all apart from J. P. Botterill's provide hay; corn is sold in small quantities by a firm of feed merchants at Ballsbridge Sales.

When young horses are put into strange loose-boxes for the first time they often tend to get excited and rush around. Therefore you should make quite certain that there is plenty of straw on the floor before putting a horse into a loose-box at the sales. If you put some hay in the corner, fill the water manger and put a feed in at the same time, the horse will probably tend to settle down sooner. Water buckets, muck forks, and yard brushes etc. are usually provided on request by most of the sales companies. It is advisable to leave the animals alone until they have settled down and recovered from their journey, before you clean them up and remove any sweat marks.

Lot numbers, which should be fastened on to the bridle when the animal is sold, are usually obtained from the foreman's office, but if in doubt apply to the sales office. In the case of youngstock, particularly foals, the numbers look better if they are trimmed round neatly with a pair of scissors before being tied on to the bridle (see Fig. 20). A word of warning here: make sure the numbers are put on to the bridle the right side up, as nothing looks worse than upside-down numbers on a bridle. It is also a good plan to fasten a

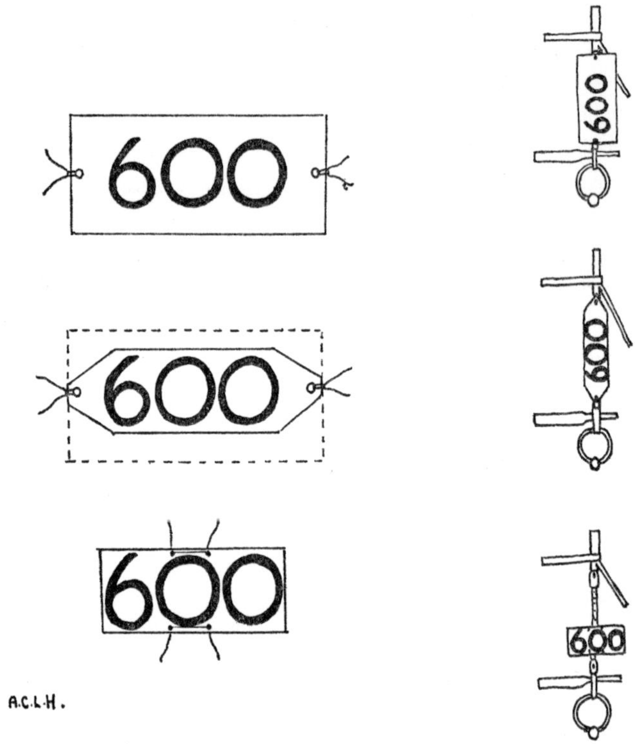

Fig. 20. Diagrams showing methods of trimming lot numbers so that they fix neatly on to a bridle or headcollar. Note especially which way up the lot numbers should be

lot number to the horse's bridle or headcollar every time he is taken out of his box, either to show him to prospective buyers or just for exercise. In this way he becomes a walking advertisement, not just an unknown horse.

Organise your staff so that there is somebody with the horses from at least 7 o'clock in the morning until about 6 o'clock at night, or whenever it gets dark. The agencies tend to come round early in the morning to look at lots in which they are interested and nothing is more infuriating for them than to find nobody with a horse, particularly if there is no headcollar on the animal either. So if animals are left unattended for any reason at least leave headcollars on so that prospective buyers can examine them in your absence.

Where more than one man is employed at the sales meal times should be staggered so that a constant watch can be maintained. Exercising is better done first thing in the morning before the sale starts and while everything is quiet, so the earlier this takes place the better.

It is probably a good idea not to muck the loose-boxes out every day but keep them clean at all times by picking up any droppings during the day and adding a clean bale of straw each morning as necessary – horses always look their best when standing in a deep bed of straw and anyway deep litter is warmer. Make sure the front of the loose-boxes is kept absolutely clean – this creates a good impression when the buyers come round. Gravel paths must be constantly raked to remove straw and hoof prints, and concrete should be swept clean.

As far as possible do not alter the horse's diet when you bring him to the sales – he must be kept going right up to the very last moment if he is to look his best in the sale ring. If you have achieved a good shine on his coat by means of feeding you need not worry too much about grooming him before the sale. Simply get any mud, muck or stains off his coat with a dandy brush, then brush him over with a slightly damp body brush at the sale, this will have the effect of removing any surface grease and dust from his coat. As soon as the coat is dry, which unless your brush was too wet normally only takes a few minutes, rub the coat over with a coarse stable rubber – you may be surprised what an effect you can achieve with so little effort. White markings on the horse's legs can be made Daz white by rubbing them over with a block of white dog chalk, which can be obtained from most pet stores. Hooves should be oiled to complete the picture.

Horses whose tails have been pulled must wear tail bandages, at least at night, to keep the hairs at the top, of the tail looking neat and tidy. Foals' tails can be improved in appearance if a tail bandage is applied for a short time each day. It is not customary to plait the manes or tails of Thoroughbred foals and yearlings at bloodstock sales. The only possible exception to this rule might be when a yearling rubs his mane at the last minute and nothing else can be done to make it look tidy on the day.

Make sure all relevant documents have been lodged with the auctioneers before the sale starts. It is customary to sell yearlings and foals without veterinary certificates, so in their case you will only need to lodge the vaccination certificates and fix your reserve prices.

Brief each man with all the up-to-date details of the animal he is about to lead round, so that he will be better able to give intelligent answers to the prospective buyers. But make sure you are readily available yourself to answer questions should anyone be interested in a particular individual.

It is better not to bring foals and yearlings out too early otherwise they will be tired by the time they get into the actual sale ring. But at the same time leave yourself enough time so that the buyers can pull the animals out while they are walking round and examine them at their leisure.

When your animals are being sold you have the right to go up into the rostrum with the auctioneer if you so wish. At some local auctions this is necessary so that you can instruct the auctioneer when to put your animal 'on the market', but at most of the major sales, the reserve price is fixed before the sale and cannot be altered when the animal is actually in the ring. As a general rule reserves should be kept as low as possible, since many buyers will not 'have a go' until an animal is 'on the market'; high reserves only tend to put them off. In most cases auctioneers will run animals up to the stated reserve in the hope that somebody will jump in at the last moment and buy the horse. At the same time it is not a good policy to fix too low a reserve on an animal, as this can have a psychologically bad effect on the buyers and the animal may never get up to his true value.

Whether your animal has been sold or not, it is good manners to find the auctioneer afterwards and thank him for at least trying to sell the horse for you.

10

NOTES ON BUYING THOROUGHBRED FOALS FOR RESALE AS YEARLINGS

A lot of money has been both made and lost by buying foals with the specific intention of selling them as yearlings. It is obviously much easier to make a profit in a rising market, particularly when times are good, but with careful planning and a certain amount of luck, money can be made, even in a bad market.

Assuming you have a place in a major yearling sale the first and most important factor is to study the market requirements closely. These can and often do vary from year to year, but basically the demand is usually for a well-bred, well-grown, good-looking individual by a fashionable sire and coming from an equally fashionable female line, preferably closely related to a good winner.

Anyone who spends a day or two at the yearling sales can tell you what is fashionable at the moment, but the art lies in being able to predict what is likely to be fashionable and therefore in demand the following year. Some people are lucky enough to have a real flair for this, as well as a good eye for a horse, but there are several factors which if taken into consideration can at least improve your chances of 'hitting the jackpot'.

First of all a certain amount of homework must be done.

Go through all the good two-year-olds which have been racing both at home and abroad, particularly those in your own country – they have the biggest impact on the home market. Decide in your own mind which you think will train on and make top-class three-year-olds and make a note of their families (female line).

Keep all the yearling sale returns. These can be found in the British Isles in such publications as *The Sporting Life, Horse and Hound* and *The Irish Field.*

Buy a copy of *The Statistical Record Return of Mares,* which covers the year relating to the yearling sales, and a current *Timeform Black Book.*

When you receive your foal sale catalogue, go through it and for each animal in which you are interested, note:

(1) What the foal's yearling brother or sister (half or own) is by, how much it made if it was sold, and most important of all, who bought it – the trainer will have some significance in assessing its chances of winning the next year. If the yearling was not sold, you can note the breeder and this may give you some clue to the likely trainer if it is being retained by the breeder. If the second dam is young enough to have a yearling too, this should also be assessed and traced.

(2) Next, look the current two-year-old's form up in the *Black Book.* If the animal is not a winner, you will have to make your own assessment of its chances of winning at three years. Obviously the more black type which appears in the pedigree *after* you have bought your foal the better your chance of making a profit, particularly if they are Pattern race winners.

Make a list of the foals in which you are interested as you go along.

(3) Check the sires and mentally give them a 'fashion rating', eliminating all those from your list which you know or believe will not be fashionable.

If you do not wish to gamble on the possible likely demand for a particular sire's progeny, it is safer to buy foals by an established sire or a first season sire – there is always a good demand for the latter.

If on the other hand you like to have a gamble, then buy foals by sires who are going to have their first runners the next year. In this case you will just have to play a hunch and keep your fingers crossed. If the sire is a failure you will be left with 'a dead duck'; but if he is a success then you will be nearly certain to make a good profit. You can reduce the odds slightly by studying the list of mares the stallion in question received in his first season and assessing the chances of the live foals becoming two-year-old winners. Again, check which trainers have the yearlings. The number of live foals is also a major factor.

(4) Finally go through your remaining foals and see which, if any, are closely related to an individual you think might make a top-class three-year-old, or from your researches among the yearlings, a top-class two-year-old the following year

You are now armed with a list of possible candidates – on paper. The next task is to make a list of the lot numbers in which you are interested, put away your catalogue so that you are not influenced one way or the other by the pedigree, and examine each foal as an individual, for conformation and soundness. This must be done systematically, otherwise it is very easy to miss some important factor – foals are always sold as they stand, without any warranty. It is therefore not a bad idea to devise your own inspection chart and simply tick off the following points as you go along:

Lot. No:
General impression:
Height: Bone:
Head: Mouth:

Genital organs:
Legs:
Feet: Action:
Temperament:
Heart: Eyes:

First impressions are very important and it is quite useful to devise your own rating system – say *, ** or *** – according to how much you like the individual foal. Yearling buyers will also be influenced by the first impression more than anything else, as they will probably not have time to look at the animal for long enough to see if it grows on them – so instant appeal is very important.

The height of the foal must be related to the date of birth and the way the foal has been done. Potential size at maturity must also be gauged. A foal which is higher behind than in front will usually grow on. Leggy foals will also tend to grow well, but in their case the depth of girth should also be noted – animals with no depth of girth appear leggy but will never mature into handsome individuals. Bone is the circumference of the cannon measured immediately below the knee. When viewed from the side, the measurement below the knee should not strike you as appreciably less than the measurement further down the cannon, otherwise the animal is what is described as tied in below the knee, which amounts to a structural weakness at this point. Good bone is desirable but an animal which is slightly light of bone should not be written off entirely, if all other factors are in proportion; that is, if it is not also heavy topped, which would make it difficult to train. An enormous amount of bone is often coupled with a degree of coarseness in the individual and light bone with quality. Colts usually have more bone than fillies. However, as with all other factors, any market fashions of the moment should be noted.

A pretty head is desirable. The head is usually the first thing a buyer will look at when he inspects the animal;

therefore a large plain head with a Roman nose is not desirable, even though it doesn't stop an animal winning races. Check the mouth: the front teeth should meet exactly, although a mouth which is slightly over or undershot, so long as the teeth are in contact, does not matter too much. If the teeth are not in contact (parrot-mouthed), the horse will have difficulty grazing and will consequently not be easy to feed. There is little demand for animals with parrot mouths.

Check the genital organs of colts – both testicles should be visible. If you can only see one or none at all and you are interested in buying the individual, it would be wise to ask one of the veterinary surgeons in attendance at the sale to check the animal for you. He can then examine the heart and eyes at the same time and double check everything else, just in case you have missed something important.

Examine the foal's legs and joints, preferably while he is still in the loose-box, and then ask the groom to walk the animal out for you. Make sure the foal's action is straight and that it does not turn its toes in or out and that it can really cover the ground. If one of its feet does turn slightly, this could be straightened by your blacksmith quite easily over a period of a few months, providing that it is only the foot that is turned and not some structural defect higher up the leg, causing the whole or part of the leg to twist. Check also that the animal does not have curbs, splints or round joints, and is absolutely sound. Look at the feet, making sure that they are a good shape and a pair – nobody will want a yearling with boxy or odd feet.

Having inspected the foals in which you are interested, you will then have to sit down and sift through all the information and draw up a final list. If you have a limited number of places in the yearling sales or limited accommodation at home, it will create fewer problems if you buy foals of the same sex. Colts tend to sell slightly better as yearlings but fillies represent less of a financial loss should they injure themselves, as they can always be sold as potential broodmares, especially if they are well bred. There is little demand for an unsound colt, however well bred he is. Most people these days look for potential stallions which have above-average racecourse ability as well as good looks.

When bidding for your foals it pays to stand in a position where you can see the animals walking round easily and preferably where you can have a last look at them before they actually come into the ring.

Having bought an animal, it will pay you to have a word with the breeder or his groom and ascertain as much as you can about the animal's diet, etc. If convenient, ask them to give the foal a final feed before they leave, which will relieve you of the trouble of feeding the animal. But yours is the final responsibility, so check that it has been done. Bring some adjustable small yearling headcollars with you and put these on the foals as you buy them – headcollars are not included in the sale price.

You should also arrange insurance as soon as possible, preferably beforehand so that any animals bought by you are covered from the drop of the hammer. You should collect any vaccination certificates from the auctioneers and check when the animals have to be moved from the sale paddocks and then make your arrangements accordingly. If you are buying a batch of animals and some have to be moved before you are ready to transport them all, or indeed have made your final purchases, the auctioneers will usually help you find alternative accommodation if you have no contacts yourself.

Most foals which have been prepared for sale will have been kept on a high plane of nutrition and may have been kept inside for a week or so before the sales; therefore great care must be taken when turning them out for the first time. Choose a well-fenced, flat paddock and turn two of the foals out together, waiting until they have settled before you add another one. If you were to turn all the foals out at the same time, they would all gallop round flat out, and one or more might well finish up by going through or over a fence.

Do not attempt to remove the headcollars when you turn the foals out, otherwise you may never be able to catch them again. They should soon settle down in their new surroundings and become used to the people who are looking after them each day.

11

CASTRATION

Castration is the removal of the male sex organs, the testicles, in order to render the animal more docile and easy to handle. It also removes both the primary and secondary male characteristics, to a greater or lesser extent, depending on the age of the animal when the operation is carried out. When colts are castrated as yearlings they do not develop a crest, the typical voice of the stallion or his naturally aggressive nature. Castration renders male horses safe to ride and turn out with mares; in some cases it also makes the animal a better jumper.

The castrated male horse is known as a gelding. The operation must be performed by your veterinary surgeon, and cannot be performed until both testicles have decended into the scrotum. In the majority of cases the testicles are in the scrotum at birth, but sometimes they do not appear until the animal is about 18 months old. Occasionally animals are found with only one testicle descended, and more rarely still with neither descended. These animals are known as rigs, or more correctly cryptorchids; those with one testicle down being referred to as monorchids.

Castration is best carried out in the spring once the frosty nights have gone and before the flies come out, or in the autumn when the flies have gone and before the frosts have started. The horse should be starved the night before the operation, particularly if it is going to be done under a general anaesthetic. If the veterinary surgeon is coming to do the operation on your own stud and is going to use a general anaesthetic you must choose a flat area of paddock which is free from obstacles and well away from a fence so that the horse cannot injure himself when he goes down and when he comes round.

Castration may be performed under either: general anaesthetic, with the animal on the ground; or local anaesthetic, with the animal standing.

Many owners prefer the operation to be done under a local anaesthetic as there is possibly less risk of injury to the colt when he is not cast, but animals must be well handled and respond to discipline if there is not to be a greater risk both to your colt and the veterinary surgeon. A general anaesthetic is essential in any case of 'rigs' and also where a hernia into the scrotal sac is suspected. If castration was carried out under these circumstances in the standing position, the intestines would drop through the cut and the animal might well have to be destroyed.

When the operation has been performed under a general anaesthetic, somebody should stay with the animal until he has come round. A lungeing rein should be attached to the headcollar so that the horse can be controlled when he comes to, but care should be taken that the attendant does not get too close to the horse in the early stages otherwise the horse might fall on top of him in his efforts to get up. Once the horse is standing he can be walked around for several minutes and then loosed in the middle of the paddock.

Exercise is of great importance following castration as it helps to reduce any swelling in the hind legs and scrotum. Horses which for any reason have to be housed after castration, *must be lunged every day* and must be kept on very clean bedding. Should the scrotum start to swell unduly your veterinary surgeon should be called in immediately as some infection may be present. It is a good idea to take the animal's temperature every day for about ten days so that any slight rise can be noted before signs of a mild infection or general septicaemia become apparent.

The castration of rigs (cryptorchids) may involve a major abdominal operation, as the undescended testicle sometimes lies inside the body cavity.

Uncastrated rigs are usually most undesirable animals; some are capable of getting mares in-foal and exhibit many male characteristics, including screaming and striking with their front legs.

Where valuable pedigree stock is concerned it is often better to leave them uncastrated until they have proved their worth (or otherwise) on the racecourse or in the show-ring. Nothing is more infuriating than to own a champion gelding, which would have been worth many hundreds of pounds more as a stallion. Therefore as a rule all pedigree stock of sufficient merit should be left uncastrated until the end of their yearling year, or in the case of 'flat' racehorses, until at least the end of their two-year-old season. It is usual to castrate all 'National Hunt' colts as yearlings. All other stock should be castrated as foals or yearlings.

12

MANAGEMENT OF THE STORE HORSE

A store horse is any horse which, for economic reasons, does not need to be broken before it is completely mature. It can therefore be kept (stored) until it is at least three years old before it is broken and ridden.

The degree of maturity attained by any horse at a given age is mainly due to two factors: heredity and environment.

Most horses do not become mature enough to ride until they are at least three years old and therefore should not be broken until this age. Ridden classes at shows are confined to horses of four years and over for this reason.

As mentioned in Chapter 3, the epiphyses or growth plates on the long bones in the horse's legs do not close finally until the animal is at least 18 months old. The epiphysis at the end of the cannon bone – immediately above the fetlock joint – is usually closed by the time the foal is about 9–12 months old, whereas the epiphysis at the end of the radius – immediately above the knee joint – does not close until the animal is about 2–2½ years old. When yearling buyers describe an apparently immature individual, as having 'open knees', they are referring to the extent to which the epiphysis at the bottom of the radius is closing. Since this is the last growing point to close it acts as a guide to the general maturity of any young horse. It is imposible to tell with certainty exactly when the epiphyses have closed by visual examination alone. Therefore many people now ask their veterinary surgeons to X-ray the growth plates at the end of the radius of valuable youngstock before they put them into work. It is also common practice in America, when Thoroughbred yearlings are sold at public auction, to have X-rays of the growth plates on display.

If an animal is worked before the epiphyses have closed, inflammation is produced which results in a soft swelling just above the knee joint (in yearlings), mainly on the inside of the leg (see Fig. 7). If the animal remains in work, the epiphysis gets wider, the inflammation spreads, fibrous tissue forms and bony enlargements appear. In all probability the animal will not go lame initially, but some pain will be shown if the swellings are pressed. In the long term, arthritis may develop, in which case the animal will go lame.

Ideally, horses should be allowed to mature slowly and not be rushed at any period of their development. The very worst thing one can do to a foal, in fact, is to prepare it for a sale. A store foal or yearling which is not going to be broken until it is about four years old should be allowed to mature naturally at an even rate of growth. This will eliminate the harmful side effects which can be created by an unnaturally accelerated rate of growth before the young horse's skeletal development is complete. For ease of management store colts are usually gelded as yearlings, but some people prefer to geld all their colt foals (see Chapter 11).

Store horses are usually marketed as unbroken three-year-olds, when they have normally attained their greatest value as young horses. As this entails keeping them for some 3–4 years, the overall costs of production must be kept within reasonable limits. Sufficient cheap grazing must be available throughout the year, otherwise the production of store horses as such is better not attempted. High labour and feeding costs would make for a very uneconomic enterprise.

Basically there are three main methods of keeping young store horses:

Running them out day and night throughout the year

Most horses stand dry cold weather but few do well when it is both cold and wet. Therefore, where possible, field shelters should be provided. These should be sited with their backs to the prevailing wind, on a very well-drained, dry base. They can be equipped with hay racks and feed troughs so that the horses can be fed inside the shelter during the winter months, but all field shelters should have virtually open fronts so that any horse which is being bullied by its companions has a readily-available means of escape, without which serious accidents can happen, especially at feeding time.

When horses are to remain out during the winter, the dryest and most sheltered fields on the farm should be chosen. Horses tend to poach land badly when the weather is wet, so cracked heels must be especially looked for and, if they occur, treated immediately with a liberal application of sulphonilamide and liquid paraffin, warmed and mixed in equal parts, otherwise severe inflammation and lameness (mud fever) may develop. Young horses must be kept in reasonable condition throughout the year so that they can grow and mature normally. A careful watch must therefore be kept on the amount and quality of grass available. The moment the value or quantity of grass declines, supplementary feeding must be given. The easiest way to do this is to use a Land-Rover to cart bales of hay or bags of nuts out to the fields. Land-Rovers are safer than tractors and trailers, as there is always the possibility that one of the horses could get pushed over the trailer bar by its companions. As a rough guide allow one bale of hay between 2–4 animals and ½ cwt. nuts between 8 animals, according to the amount of grass.

Drop the hay and nuts off in piles across the field, allowing at least three more piles of each than animals in the paddock to prevent fighting. Choose the cleanest and most sheltered areas, giving sufficient room between the various piles and the nearest fence or hedge so that bullied animals do not find themselves trapped. Alternatively the food can be placed in a field shelter, as already described, or in metal food troughs fixed on to the fence (see Fig. 21). Salt licks or lumps of rock salt should be made available at all times throughout the year. As far as possible, animals of a similar size should be placed together. It would be asking for trouble to run weaned foals with strong three-year-olds, particularly when they are to be fed together. You will be able to judge when to stop feeding the horses in the spring as they will stop cleaning up the hay.

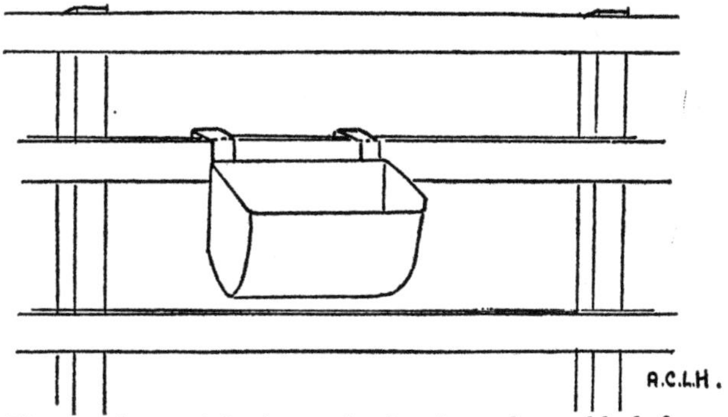

Fig. 21. A metal feed trough placed on the paddock fence

During frosty weather the water troughs must be inspected at least twice a day and the ice broken when necessary so the horses don't go short of water. Make sure that any automatic filling device still works – external pipes are better lagged before the hard weather comes. Chalk can be placed around water troughs and gate-ways to help reduce poaching.

Yarding the horses during the winter months

Deep litter bedding should be used and the covered yard mucked out in the spring after the horses have been turned out to grass. A building with a sufficiently high roof should be used, as a low-roofed building would have to be mucked out during the winter when the level of litter got too high.

Horses, unlike ruminants, do not thrive on a high-fibre

diet, so yarded horses fed nothing but hay, unless it is very good quality, will usually lose weight. Feed troughs should be sited down side walls of the yard allowing plenty of feed space per horse. Salt licks should be provided, and a suitable water trough, which must be cleaned out regularly. Adequate ventilation is essential, as stuffy conditions can lead to respiratory troubles.

When horses are yarded together it is even more important not to mix the sizes too drastically, otherwise accidents might happen.

Yarding has the advantage that the horses are not out on the land when it is very wet and therefore will not be cutting the fields up at that time of the year, or developing cracked heels. Horses' feet must be attended to regularly when the animals are running out at grass all the year round or yarded on deep litter. As there is no natural wear of the hoof wall on soft ground, the hoof soon starts to grow out of shape if neglected, and this can be difficult to correct later. Also, an earlier bite will be obtained in the spring if the paddocks remain empty all winter.

Kept in loose-boxes during the winter, turned out each day

If the loose-boxes are mucked out each day the labour requirements of this system would be too high to justify its use for store horses; that is unless the owner looks on the enterprise as a hobby and does all the work himself. Otherwise deep litter must be used.

By bringing the horses in every day they are being handled regularly and therefore don't get the chance to become half wild, which can easily happen, especially if they are running at grass all year round. The other advantage of this system is that individual horses get a known amount of food, and greedy horses cannot get fat at the expense of their more fastidious companions.

Even if animals are kept in fields all their lives, some loose-box accommodation must be available. Sick or injured animals may need to be brought inside at any time. Most young horses are easier to cope with if they are housed to have their feet trimmed, or for worming. This does also mean that the young horses are handled from time to time.

Yearlings, two-year-olds and three-year-olds, which are of the correct type and also sufficiently good looking, can be shown throughout the summer months. If they are taken to a few shows each year you will find that the young horses are virtually broken by the time they are three years old. As a rough guide, for hunter classes, colt foals should measure 14.2 h.h. by the turn of the year and fillies some 2 in. less, if they are going to have much chance at county show level. By the time they are two years old they should be about 16.0 h.h.

Some of the bloodstock sales companies organise special sales for stock with potential as National Hunt horses (horses likely to win races over hurdles and fences). They will usually accept half-bred horses as well as Thoroughbreds, so long as they have some potential as racehorses, even if it is only as point-to-pointers. The best sales, financially, are held about the beginning of June.

The main sales which fall into this bracket are:

J. P. Botterill's – Ascot sale, at the beginning of June

Doncaster Bloodstock Sales – Great Annual Spring Sale, mid-May

Doncaster Bloodstock Sales – Great Annual York Race Sales, mid-August

(These two sales at Doncaster are usually preceded by in-hand show classes for unraced three and four-year-olds likely to make high-class National Hunt horses.)

Ballsbridge Sales – Derby sale, at the end of June

Ballsbridge Sales – Dublin Horse Show Sales, for owners of exhibited horses, not necessarily with any potential as race-horses but rather as good riding horses, hunters, show jumpers, etc.

(Many of the breed societies also hold their own sales each year for registered stock.)

In England most horses are now sold subject to re-examination by a veterinary surgeon on behalf of the purchaser, not the vendor. This is in line with the recommendation made by both the Royal College of Veterinary Surgeons and the British Veterinary Association. Veterinary surgeons are in attendance at all sales for this purpose. The vendor can have a horse examined at home, to satisfy himself that the horse he is selling is sound, but now the examining veterinary surgeon will not usually give a certificate for that horse to be read out at the sale.

In cases where an animal has had a minor accident which has left a slight blemish, it is better to be guided by your veterinary surgeon as to whether this should be mentioned. If the blemish is virtually unnoticeable, in most cases it is better to say nothing. In the case of yearlings and foals, which are normally sold as they stand, it might be better to have an X-ray available and a veterinary certificate for any noticeable injury than to leave the buyers guessing as to the extent of the damage and its possible consequences.

When selling two-year-olds and older horses it is customary to give a short description and history of the animal on the following lines:

This . . . is unbroken but well handled.

Has been driven in long-reins.

Has been carefully mouthed and backed but not ridden away.

Ridden about the farm.

Broken last winter and turned away to mature.

Guaranteed untried (broken stock only).

If one is going to maintain a good name, it is never sensible to sell a horse with a known fault without mentioning it. Honesty is the best policy.

13

THE ART OF LOADING AND TRAVELLING YOUNG HORSES BY ROAD

These days all horses and ponies travel by road some time during their lives, and those bred specifically for showing, hunting or racing will probably spend a large portion of their time on the road. They should therefore become accustomed to loading and travelling from an early age.

It is essential to have some facilities for loading and unloading horses on a stud. Even if you only own a trailer yourself, you should make provision for horse-boxes, as you will never know when one will visit your stud to collect a horse. All trailers are fitted with low-loading ramps, so the only thing necessary in their case is to build a loading ramp without any rise in ground-level or to select a quiet spot where you can run the trailer tight up against one wall of a building, but make sure that the trailer really is tight up against the wall, otherwise a horse might get a leg down between the ramp and the wall and injure itself. In the case of horse-boxes, it is almost essential to build a loading ramp as shown in Fig. 22.

This should be sited in such a position on the stud that large horse-boxes have plenty of room to turn round, and the ground in front of the ramp must be solid enough to take

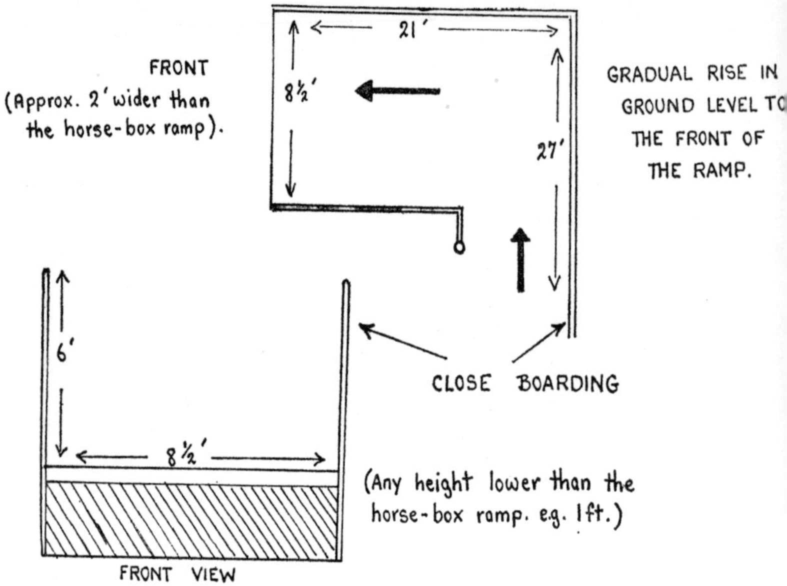

Fig. 22. Design and measurements of a loading ramp

the weight of heavy vehicles even in wet weather. Most horse-boxes have their loading ramps on the near side of the vehicle, so they must have room to pull up straight alongside the ramp on this side.

It is much easier to load horses in a quiet place away from all other animals, so loading ramps should be built out of sight of the paddocks. For this reason it is also better to close-board the shute which leads to the ramp itself. This shute should be made in such a way that the ground level gradually rises to meet the height of the ramp. The surface should be well drained and preferably grassed over, or have some other form of non-slip surface. This is particularly important when unloading, as some horses have the habit of jumping off ramps. Shutes should also be made long enough so that horses can be walked into them easily when a horse box is in position for loading, and sited so that horses do not have to pass the vehicle.

Before travelling mares with foals at foot, the following points should be borne in mind:

Where the foal has access to the front of the trailer or is to be travelled in a groom's compartment, the groom's door must be securely locked from the outside to prevent any chance of the foal opening the door from the inside by leaning on the handle.

Windows which are low enough to be reached by the foal must be fitted with a wire mesh guard to prevent any chance of the foal injuring itself.

Where mares are to be travelled with foals at foot, the centre partition should be removed from two-horse trailers to convert them into a single loose-box (the same applies to horse-boxes). (See Fig. 23.) In the case of trailers it is

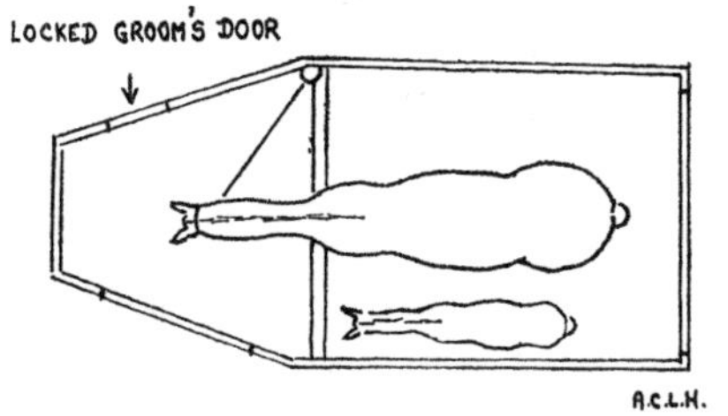

Fig. 23. Two-horse trailer with centre partition removed to accommodate mare with foal at foot

essential that a metal breast bar is fitted, so that the mare is not thrown to the front of the trailer whenever the brakes are applied. The mare must be tied up but the foal should be travelled loose, preferably without a headcollar, so that it cannot get caught up on anything during the journey. The breast bar can be padded with foam rubber for extra protection when the foal walks underneath. As foals mostly travel lying down, the floor should be well bedded down with straw.

If hay-nets are used in horse-boxes or trailers, they must be tied up as tightly and as high as possible, preferably using two strings. They must be checked periodically during the journey as if they drop down a horse could easily get a leg stuck in the net.

Mares should not be travelled in stalls or single-horse trailers with foals at foot as there is insufficient room, unless the foal is placed in the front of the trailer, with a hinged solid partition between it and the mare, or in the groom's compartment of a horse-box (see Fig. 24). If a foal is placed

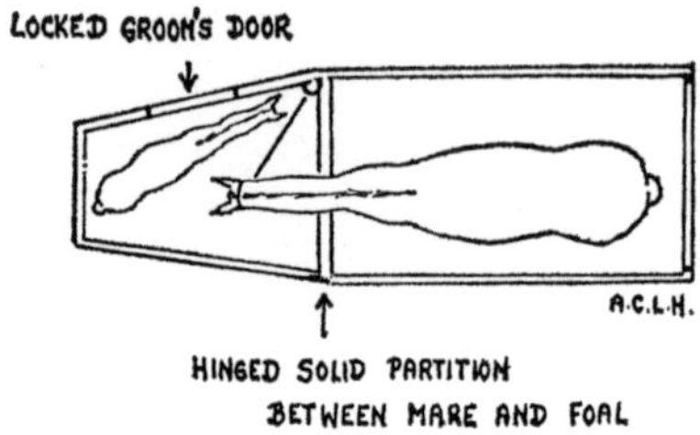

Fig. 24. Single-horse trailer with hinged partition to accommodate mare with foal at foot

in the groom's compartment, no other horses should be travelled in such a position that they can bite the foal. In the case of young foals, under a month old, a stop must be made at least once every two hours, to allow them to suck the mare. Older foals can go a little longer without milk.

When travelling mares with foals at foot, both trailer ramp top doors should be kept closed in cold weather to help eliminate any draughts, and in the case of older foals to prevent any chance of their jumping out during the journey. However, in hot weather some ventilation must be provided. To achieve this the left-hand top door should be fastened back or removed and replaced with a medium-fine mesh wire-netting (see Fig. 25). This should be tied on to the door hinges and fastenings and will provide sufficient air, at the same time preventing any chance of the foal jumping out. The off side top door should be kept closed to prevent the mare from wanting to turn round and also cut out the sight of overtaking traffic.

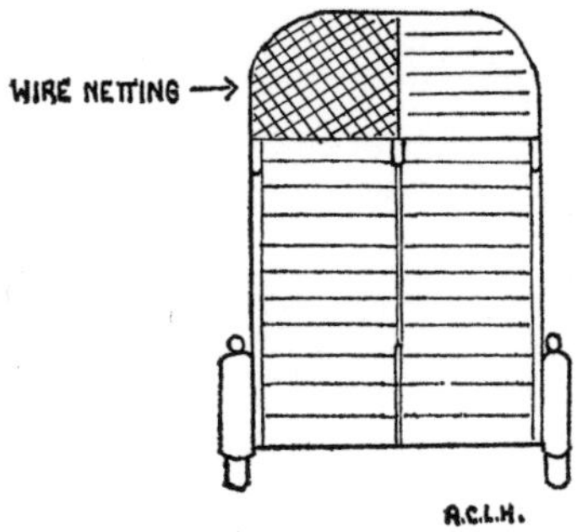

Fig. 25. Using fine mesh wire-netting to achieve ventilation in the trailer

Loading horses into trailers and horse-boxes can be a very difficult task unless a little horse-sense and tact is applied. Then what seemed previously to be an impossible task suddenly becomes easy.

Most horses are willing to co-operate if they feel they are in a confined space with little room or opportunity to play up. To this end a loading ramp as already described is an almost essential piece of equipment to facilitate the loading of animals into horse-boxes. As far as trailers are concerned, the great thing to remember is that unless the animal is known to be easy to load, one should not try to load in the middle of an open space, but should site the trailer close to a wall or in a confined gateway. Leaving the front ramp top door open lets more light into the trailer and sometimes helps a nervous horse.

Horses do not like to step on to a surface which is not absolutely solid. Therefore ramps should be checked to make sure that they are level and that there is no movement. If the ramp is not on level ground, pieces of flat stone or wood should be placed under it until there is no possible movement when someone walks on the ramp.

Young or inexperienced horses can be encouraged to walk on to a ramp more easily if the trailer or horse-box, its ramp and the immediate area in front of the ramp are covered with straw. This makes it look less frightening. The ramp itself should have the minimum slope – hence the use of a loading ramp, as described on page 136. As far as possible, inexperienced or nervous horses should be given the maximum width of stall to walk into. To achieve this, the partition should be moved to one side, and replaced once the horse is loaded (see Fig. 26). Then the breaching strap should be

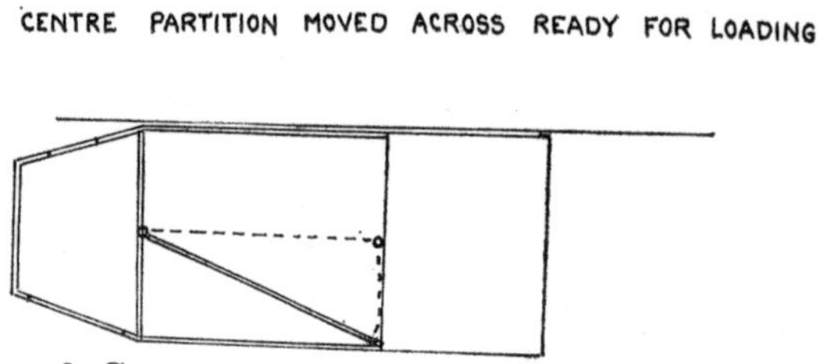

Fig. 26. Centre partition moved across ready for loading

fastened before the centre partition is secured, or the back ramp put up, otherwise the horse might become frightened and attempt to run out backwards as the partition is moved.

When loading horses into vehicles and trailers, the handler should walk by the horse's head and on no account should he turn to look at the horse, as nothing is calculated to stop a horse quicker than to turn round and look him in the face. The handler should not duck under the bar until the breaching strap has been fastened by an assistant or the back ramp has been put up, otherwise if the horse decides to reverse out of the trailer or horse-box suddenly, the handler will be compelled to let the horse go. For the same reason, the horse should not be tied up until his exit is securely barred, otherwise he would almost certainly break his head-collar.

To counteract the natural camber of the road, when only one horse is being travelled in a two-horse trailer it should be loaded into the off-side stall. When more than one horse is to be travelled at the same time, the heavier one should be put on the off side. In the case of mares with foals at foot, the foal is left loose but the mare should be tied to the off side ring in the horse-box or trailer. When tying horses, the rope should be secured with a slip knot which can be untied easily when necessary. When two or more horses are being travelled, head boards should be fitted, or the horses secured so that they cannot touch noses, otherwise they might start fighting during the journey. To prevent the headcollar from being broken in the event of a horse pulling back suddenly, some people fasten a loop of baler twine to the ring in the vehicle and tie the rope to the twine.

Horses which are reluctant to load are often encouraged to do so if a lungeing rein is fastened to the far side of the trailer, passed round the hind-quarters just above the hocks and pulled tightly. This method requires a minimum amount of labour. (See Fig. 27.)

Mares with foals at foot, particularly young foals, can be encouraged to load easily if the bar at the front of the trailer is removed (as in Fig. 28) to give the maximum amount of

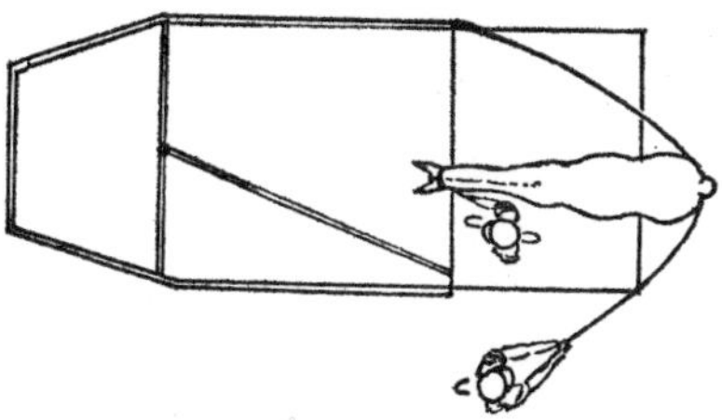

Fig. 27. A lungeing rein passed round the horse's hind-quarters just above the hocks will often make loading easier

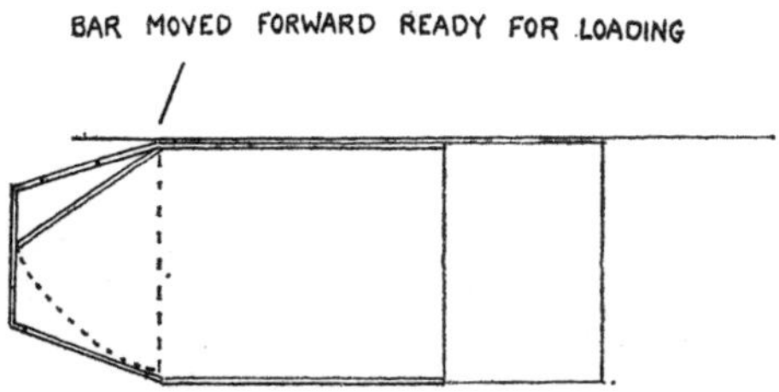

Fig. 28. Bar at the front of the trailer removed to give the mare and her foal the maximum amount of floor space when loading

floor space and the foal is loaded before the mare and taken up to the front of the trailer. If the mare is kept as close to her foal as possible all the time she will usually follow it

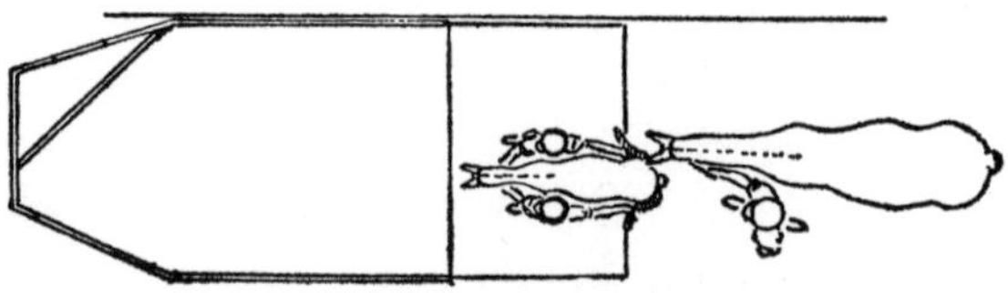

Fig. 29 Load the foal first, and keep the mare close to her foal

into the trailer without any trouble. (See Fig. 29.) The bar is then replaced after the ramp has been put up.

Horses which are being shown or taken to sales must have their tails protected, otherwise they will rub the top of their docks during the journey. This will spoil their appearance and make the tail impossible to plait should this be required. Therefore it is customary to put tail bandages on horses before travelling. In order to keep tails clean it is also a good idea to cut a leg out of an old pair of tights and put the tail inside, securing it with a tail bandage put on over the top in the normal way. For extra protection in cases where horses are accustomed to wearing a roller, a tail guard can also be used. Coconut matting fitted to the back ramps of trailers and horse boxes, as well as making the ramp more inviting for the horse to walk on to, provides a considerable amount of protection to its tail, hind-quarters and hocks during a journey. Further protection can be given by using hock boots.

Many trailers and horse-boxes have a slight space between the bottom of the partition and the floor. While travelling, horses usually keep their balance by standing with their legs apart, and so will often stand with their pasterns rubbing up against the bottom of the partition. Because of this, it is often wiser to apply stable bandages to horse's legs before loading; where there is a solid partition in front of the horse knee pads should also be worn.

Remember that it is just as important to get a trailer serviced regularly, as it is a horse-box. By doing so you will ensure that the braking system and hitch are maintained in good working order. The condition of the floor should also be checked at regular intervals to make sure there are no signs of excessive wear or rot. A strong safety chain should be fitted to all trailers.

As far as youngstock are concerned, trailers fitted with front unloading ramps are preferable as these do not teach young horses the annoying habit of running backwards out of trailers, immediately after loading. Horses can also have loading practice. With the centre partition removed, they can be walked up one ramp, through the trailer and down the next, then the centre partition can be replaced and the procedure repeated, until the young horse is quite used to the idea.

When driving a vehicle which is towing a loaded trailer great care must be taken not to corner too sharply or slow down suddenly. The horses get a smoother ride if the gears are used as much as possible rather than the brakes alone.

14

PREPARATION FOR SHOWING YOUNGSTOCK IN-HAND

It you are going to show your youngster in-hand, it is essential that you should decide at the beginning of the year exactly how many shows you intend to enter and stick to this number. Otherwise showing can become a very expensive game these days, even if you are lucky enough to win prizes most of the time. It is also essential for the same reason, as well as to reduce the risk of infection, that you should have your own transport.

Bear in mind that the closing date for entries, particularly for the big shows, is often many months before the actual date of the show.

If you are not sure which shows to patronise, or indeed where to apply for the schedules, the best thing you can do is to buy the special show numbers of *Horse and Hound,* which usually come out at the beginning of March. These will give you comprehensive lists of almost every show to be held during the year, together with the secretaries' names and addresses.

Small shows will very often accept entries on the actual day of the show. However, this must be checked, as not all shows offer this concession and nothing would be more disappointing than to arrive at the show only to find that you cannot enter.

At some of the larger shows you can hire a sleeping cubicle for yourself and/or your staff. These are sometimes provided with a bed and mattress but at other shows they contain nothing. Most people, these days, sleep in their horse-box or trailer, which keeps expenses down to a minimum. Remember to pack plenty of padlocks so that you can lock up all your equipment. Tack should be kept in a large well-locked box inside a locked vehicle or building.

It is a very good idea to keep a record of every show attended, the judge, horse shown and result. In this way you will be able to see at a glance in future years which judge likes which type of horse. This information is invaluable when making up the show entries at the beginning of the year.

There is no need to school your youngster especially for the show ring if you have already taught him to lead well in-hand. Apart from this, just follow the routine described in Chapter 9 to prepare a foal or yearling for sale. A certain degree of controlled high spirits is usually overlooked where foals and yearlings are concerned; indeed it is far better if a youngster is 'on its toes' as it then tends to show itself off better than an individual which is well-behaved but half-asleep.

Any horse which is going to be shown regularly must be taught to load easily into a trailer or horse-box. The art of loading horses into vehicles is described in detail in Chapter 13.

The best time to trim your horse for the show season is as the spring approaches and he starts to lose his winter coat. At this time of the year you will find that the mane and tail hair comes out much more easily, as does any surplus hair around the fetlocks and head. If you do not intend to plait your young horse's tail each time you show him you will need to pull it. This should be done over a period of time, so as not to make the tail sore, and only a few hairs removed at a time. A tail bandage should be put on to train the hairs at the top of the tail to lie flat.

The mane must also be pulled in all cases apart from native breeds, Arabians and palominos. It is almost impossible to plait a really long mane neatly. Pull the mane to about 3–4 in. long at the beginning of the show season and you will probably only need to pull it once more towards the end of the season. Unlike the tail, this should be done all at once – if done in two stages you may well get a 'step' in your otherwise straight line.

The mane should be well brushed out and a flat-topped metal comb used. Taking only a few hairs at a time, backcomb them to the required length; wrap the long hairs you are holding in your left hand round the comb and pull them out. Continue down the mane keeping a perfectly straight line along the neck. If the mane is still too thick when you have finished it can be thinned by pulling a few hairs out from the underside, but never on top. A thin mane can be shortened by using a sharp knife instead of pulling the hairs out by the roots after you have backcombed the hair. The hairs should be broken across the length of the blade as this gives a more natural appearance. On no account should a mane ever be cut with a pair of scissors, and preferably should not be trimmed over the poll.

All horses' feet should be trimmed regularly – at least once a month and more often if possible in the case of young horses. It is customary to shoe young horses (but not foals) in front with light plates before showing them. But when it is preferred to show animals in-hand unshod, it is better if their feet are not trimmed for about two weeks before the show, otherwise if the ground is hard they may well become foot sore and go lame when trotted up.

Under the rules of the Joint Measurement Scheme, annual height certificates are not issued until animals are four years old. Therefore young ponies entered in breeding classes which have a height limit are measured on the showground. The height is the measurement from the ground to the highest point of the withers when the horse is standing with its front feet together and its poll higher than the withers. An allowance is made for shoes or plates.

Horses which turn their toes in or out can have this fault rectified or improved by careful trimming and/or shoeing from an early age. The *inside* of the foot is rasped down so that it is lower than the wall on the outside of the foot in order to throw the toe *out* and the *outside* is rasped down to bring the toe *in*. See Chapter 3, page 55. The appearance of a shod horse which toes either in or out can be improved by off-setting the clip so that it appears to be in the centre of the foot when the horse is standing in his natural position.

There are two distinct methods of keeping young horses which are going to be shown in-hand during the summer months. The first entails leaving them inside most of the time and only turning them out for a limited period each day. Accordingly, the horses are strapped daily and light summer sheets are usually worn to keep them as clean as possible, with a good shine on their coats. Under this system horses coats do tend to 'break' much earlier than is the case with horses which are only groomed at the show and are allowed to run out at grass both day and night, at least until mid-July. Under the second system a good shine is produced by feeding a linseed mash each day coupled with a quick clean-up and wipe over at the show. Care should be taken with small ponies, which are probably better without any corn, as many ponies are very prone to laminitis during the spring and summer months.

When horses are shown straight off grass they are inclined to 'dry up' on coming inside even for one night. It is therefore essential that some cut grass is fed as well as hay when they are brought in for a show, otherwise in only a matter of a few hours they will look really hollow-flanked and 'run up'. The grass must not be allowed to heat up before it is fed, otherwise it could cause colic.

It is very useful if horses can be exercised on the lunge at a show, but as the last epiphysis (at the end of the radius) does not close until the animal is at least two years old, it is advisable not to lunge yearlings. (A detailed description of how to teach horses to lunge can be found on pages 80–83 of *The Amateur Horse Breeder.*)

Very little equipment is needed for showing horses in-hand – a list of all the essentials is given on pages 151–2. The most important thing to remember is that any bridles, head-collars, or leading reins must be absolutely clean and shining. Bridles must be fitted with nosebands, otherwise the horse's face will appear bare and unfurnished. It is essential to try the bridle on the horse some weeks before the first show, so that any adjustments can be made with plenty of time to spare.

Double bridles are occasionally used for three-year-olds, but personally I much prefer to use a snaffle on young horses of any age. If ordinary bridles are used they should be stitched or fitted with studs rather than buckles. All metal parts must be shining. Stainless metal buckles are the easiest to maintain but rust can be removed from other buckles by rubbing them with very fine sandpaper.

For ease of work and a first-class appearance there is nothing to beat stainless steel for all bits. Although nickel can look nice it requires an enormous amount of work to keep it up to show standard.

Foals should be shown in brass-mounted headcollars, fitted with narrow rolled and stitched nosebands and browbands. These headcollars are usually made from leather but any serviceable material would be acceptable as long as it looks nice. Some people like to use white webbing headcollars for very young Welsh (Section A) foals. These look very good but have the disadvantage of having to be scrubbed regularly and whitened when necessary with canvas shoe cleaner, which could have disastrous results if it rains. Brass mountings should be used in preference to tin but they must be kept polished and shining.

Quiet yearling fillies may be shown in headcollars alone, without bits, but entire colts, geldings and older fillies are all safer with bits in their mouths. For leading young horses in-hand, as already explained in Chapter 9, page 109, I prefer to use a flexible rubber bit. This can quite satisfactorily be attached, by means of special leather bit attachments, to the headcollar; or better still it can be fitted to a brass-mounted in-hand bridle. Either method is equally correct but the in-hand bridle possibly looks a little neater.

Show tack should be cleaned with dark tan boot polish on the top side of the leather and only saddle soaped on the reverse side. Any new tack should be cleaned several times to darken the leather before it is used. A light tan leather gives a better final colour than a dark leather but requires harder work initially to get rid of the new yellow look. When matured it acquires a lovely reddish tone which looks well on most horses.

An extra shine can always be obtained by putting the polish on in thin layers and rubbing small areas at a time up to a shine with a little water. A glass-like surface can be produced in this manner which is useful for the front of nose-bands, browbands, etc., and also for the toes and heels of jodhpur boots. But care must be taken not to produce this shine too often, otherwise the leather may crack. As far as possible oil should not be put on show tack as once applied it will be impossible to produce a shine with boot polish ever again.

There are two main types of leading rein in common use: leather or substitute leather reins; and white web reins.

Substitute leather reins, being made from plastic, need no maintenance beyond an occasional wash down. Leather reins should never be cleaned entirely with saddle soap, otherwise if it rains the leather will become slippery and should the horse take a pull, the rein will simply slide straight through your hand. Therefore they are better cleaned in exactly the same way as the bridle, saddle soap being applied only to the reverse side of the leather. Leather reins are the most reliable as, unless badly chewed, they are unlikely to rot and break suddenly as web reins can do.

Web reins are cheaper to buy initially but last for a much shorter time, probably due to having to be scrubbed or cleaned with white canvas shoe cleaner each time they are used, as this tends to rot them.

The leading rein can be attached to the bit in one of two ways. It may be buckled to the far ring of the bit and passed through the nearside bit ring to the leader's hand; this probably gives the greatest control but is consequently harder on the horse, and also requires more horse sense and tact in handling on the part of the leader. The other method is to use a bit coupling, which is usually made from leather, but can also be in the form of a brass chain. This simply buckles or clips on to the rings of the bit and also on to the back of the noseband. This produces an even pressure on each side of the horse's mouth and also takes some of the pressure on to the front of the nose.

The leading rein can also be made with a 'Y' coupling, which has much the same effect as the bit coupling described above, or it can be attached to the off-side bit ring and passed through the near-side ring. When using a bit coupling care should be taken that the coupling fits close enough behind the horse's jaw so that he cannot grab it with his bottom teeth, as if this happened he might well rear up and come over backwards.

The appearance of a beautifully turned out horse can be absolutely spoilt by a badly turned out leader; which after hours of hard work on the horse would be a great pity, to say the least.

For showing in-hand where it is essential that you should be able to run fast when the occasion warrants, it is better to wear a pair of shoes or boots which are comfortable and easy to run in, than a pair which are fashionable. Long riding boots and high-heeled shoes are unsuitable. Jodhpur boots are ideal, and on women look smart worn with stretch beige slacks, a shirt and tie or jersey and headscarf.

Men usually wear a neat country style suit with a bowler hat; or sports jacket and cavalry twill trousers with a collar and tie, usually with a trilby hat or cap. In warm weather a cotton jacket can be worn, but however hot it gets, it is simply not done to appear in the show ring in shirt sleeves.

The following is a complete list of articles which you will need to take to the show with you, especially if you intend to stay overnight:

Headcollar
Rope
Hay net
Water bucket
Feed
Hay
Feed bucket
Scoop
Skip and mucking-out fork
Roller and tail guard
Knee pads

Grooming kit
Body brush
Dandy brush
Water brush
Curry comb
Mane comb
Hoof pick
Stable rubbers

Tail bandages
Leg bandages
Gamgee tissue

Show bridle
Lead rein
Cane

Lungeing cavesson
Lungeing whip
Lungeing rein

Plaiting equipment
Thread
Blunt needles
Scissors
Box to stand on

Dog chalk block
Hoof oil
Rag
Vaseline
Disinfectant spray
Fly repellent spray
Boot polish
Saddle soap
Brass cleaner
Cleaning cloths and sponges

Wire cage or rails
Show tickets, numbers, documents, passes, etc.
Collapsable bed
Sleeping bag and pillow
Blankets
Torch

15

SHOWING IN-HAND

When you arrive in the vicinity of the showground you will find route signs indicating livestock, horses, etc., which direct you to the correct entry gate. Be prepared to show your tickets to the man standing at the entrance gate and make sure that any livestock passes are signed where necessary. Many of the larger shows have a veterinary surgeon on the gate to check all the horses as they arrive, before they are unloaded, to make sure that no horse entering the showground has any signs of a contagious disease, such as a cough or runny nose.

Once your horse has passed the veterinary inspection make your way to the horse lines where you should find the horse foreman. He will tell you exactly where your loose-box is, and will probably have the ring numbers. At some shows you can drive down between the rows of boxes to load and unload your horses, which saves walking them from the parking area.

When you have found your loose-box check that the manger is clean. If it still contains stale food from a previous occupant this must first be cleaned out. Straw is provided by most shows, so if by any chance the box has not been bedded down, sufficient straw should be asked for. Some loose-boxes have a canvas sheet across their front. This should be tied up with string or baler twine before the horse is asked to walk into the box, otherwise it will flap round his head even if held up and might frighten him. If you have hired an extra box or cubicle for your equipment unload this before you take your horse-box or trailer to the car park.

Some loose-boxes at horse shows have very insecure door fastenings. Therefore, to be on the safe side, the bolt should be padlocked or tied up with string before you leave. Water is usually available at the end of each horse line, with a galvanised tank under the tap. Always draw your water directly from the tap – never dip your bucket into the tank, as this water is often stale and well contaminated by other people's dirty bucket bottoms.

Should the show provide hay and/or greenstuff for their livestock exhibitors, go to the forage yard as soon as possible, as they often close early in the evening and do not open again until the following morning. Forage tickets are either sent by post with the other show tickets or are available from the horse foreman and these must be presented at the forage yard. However, with the current high prices of forage many shows have stopped supplying hay and rely on exhibitors to bring their own. Greenstuff is still sometimes provided and left in the horse lines so that exhibitors can help themselves.

Evening meals are available on most showgrounds in both the members' and stockmans' canteens. The latter is usually perfectly adequate and much cheaper than the former. If you intend to sleep in your vehicle on the showground, it is definitely advisable to muck the vehicle out before it gets dark. Otherwise the operation will be doubly difficult; also the floor tends to dry off much quicker in daylight. An interior light is almost essential, otherwise you will be groping around in the dark when you go to bed. As far as the actual bed is concerned, either a camp bed can be used, or if you intend to sleep on the floor in a sleeping bag sufficient straw should be laid down to form a thick mattress. Some horse-boxes and most trailers have a corrugated floor, which is incredibly uncomfortable to lie on unless there is plenty of straw under you; then it can be very cosy.

If you are not entirely familiar with the particular show-ground it is a very good idea to have a walk round before turning in for the night. Note where the collecting ring and judging rings are so you will not have to spend valuable time in the morning looking for them. Most youngstock and breeding classes are scheduled to start at 9 a.m. or sooner.

An early start in the morning is essential as mucking out usually has to be completed by 6.30 a.m., at which time a gang of men come round with a tractor and trailer to pick it up from the front of the loose-boxes.

The loose-box should be bedded down neatly with the straw built up round the sides so that it looks tidy when the general public come round to admire the horses later in the day. Most loose-boxes at horse shows have open fronts, so if your horse is inclined to get upset when stroked, petted and prodded by many different people, or has not settled down well with either or both of the horses next to him, it is advisable to tie fine mesh wire-netting across the front of the loose-box to form a cage.

Remove any stable stains and dust from your horse's coat before breakfast, brush his mane and tail out well and plait up, if he is to be shown in a class where it is customary to have horses plaited (all classes other than mountain and moorland ponies, palominos or pure bred Arabs). Where native ponies, palominos or Arabians are shown in classes other than those listed exclusively for their particular breed they too must be plaited.

To plait a horse's mane and tail you will need the following equipment:

(i) A headcollar and rope – the horse can be tied up, or if he is only a foal or yearling he should be held by someone else;

(ii) A box to stand on – this is essential if you are small and your horse is tall;

(iii) A water brush – to dampen the mane and tail;

(iv) Some thick thread – (obtainable from most saddlers) this must be as near to the colour of the horse's mane as possible;

(v) A mane comb and scissors;

(vi) At least one blunt needle – to save time two needles can be used, the person holding the horse keeping the spare needle threaded, or a pin cushion can be filled with previously-threaded needles, but great care must be taken that none are lost in the straw.

The exact number of plaits you intend to put in your horse's mane should be determined before you start, as it is essential that every plait is made exactly the same size right down the neck. The current fashion is to make a large number of very small button plaits. These look good and greatly improve the appearance of a well-developed neck (see Plate 13).

Thread the needle and with the thread double tie a knot in the end. Stick the needle in your lapel and pass the loose end of the thread over your shoulder so that the needle does not get pulled out – there is less likelihood of the needle getting caught up in something, if it is hanging down your back.

First dampen the mane and divide the forelock from the first plait immediately behind the ear by making a parting with your comb. Divide the first plait from the second in the same way, measuring its width with your comb and relating this measurement to the length of the neck and the total number of plaits you intend to make.

Plait the hair right down to the bottom and secure the end with a knot. Bring the needle back into the base of the plait, pulling the end up to form a loop. Bring the needle back into the end of the loop and double the plait up again, making sure that any stitches on the surface of the plait are kept small so that they do not show. Next wrap the thread two or three times round the base of the plait and pass a stitch right through the thickness at the base, making sure that you catch every layer of the plait. Knot the thread off on the underside and cut it short enough not to show.*

Keeping the mane damp continue to plait down to the withers; the last plait or two can be made a little slacker to allow the horse freedom to stretch his neck without discomfort. Once the mane has been plaited up completely any short hairs which are sticking up can be pulled out – on no account should they be cut, otherwise in a week or two they will have a hedgehog appearance, and look ghastly.

If your horse's tail is not pulled, when you have finished plaiting his mane turn him round with his back to the front of the box, so that you can see what you are doing. Make quite sure that there are no knots left in his tail, then dampen it and start plaiting as high up as possible. Take three small sections of hair at the top of the tail, introduce a further strand of hairs from the right side, underneath the tail and then from the left side underneath the tail (see Fig. 30). Continue down the tail until you are about one inch from the end of the dock – unless the horse has a short tail, in which case the plait should finish correspondingly higher up. When you have reached the required length stop introducing any more hairs and simply plait the hairs in your hands down to the end of form a long pig-tail. The end should be secured in the same way as for the mane. The pig-tail can then be looped up and tucked into the plaited part of the tail and secured with a few small stitches. For a more interesting effect the pig-tail can be rolled up and secured in place with a few stitches.

The secret of good tail plaiting is to take only a very few hairs at a time, otherwise the plait will not be tight. To achieve a ridge effect down the tail, instead of passing the hairs over one another while plaiting, pass them under one another. This has the effect of making the plait stand up from the level of the rest of the tail.

*See also *Amateur Horse Breeder* A. C. Leighton Hardman, Pelham Books.

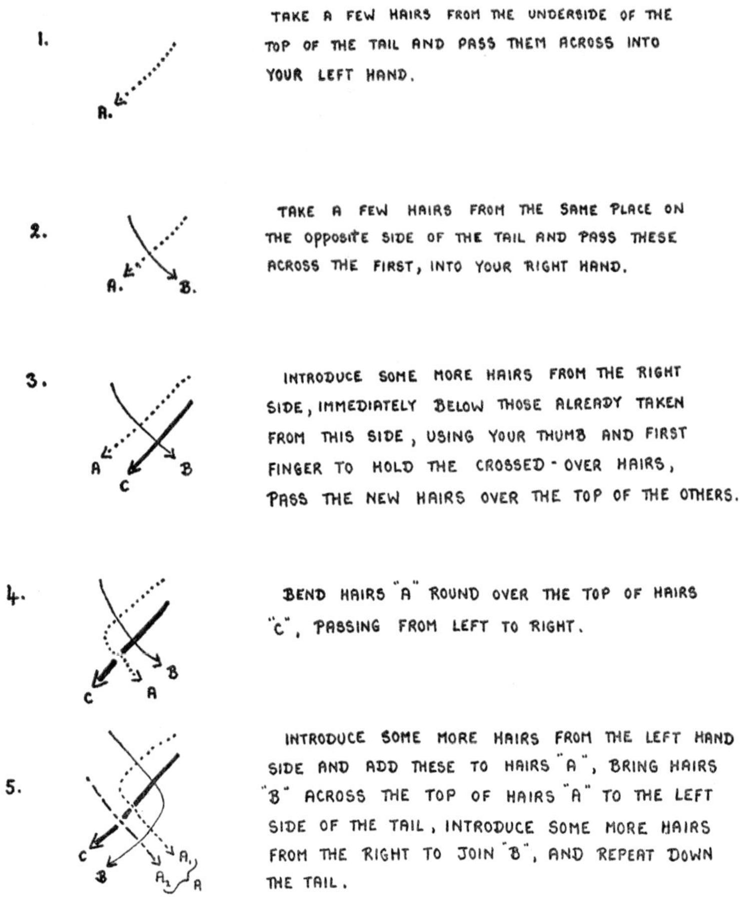

Fig. 30. Stages in plaiting a tail

A damp tail bandage should be applied as soon as the tail has been plaited. This helps to flatten the hairs and greatly improves the appearance of the plaiting. When you come to remove the tail bandage, care should be taken that it is not pulled off in the usual way, as this would probably damage the plaiting; instead, it should be carefully unwrapped.

In the case of horses shown unplaited, the mane and tail

should be brushed out thoroughly and then combed through, so that the hair is completely free from knots. In the case of mountain and moorland ponies, which are always shown unplaited in their breed classes, it is customary to plait the section of mane immediately behind the ears, so that it stays in place tidily and does not get mixed up with the forelock. The plait is left as a narrow pig-tail and not turned up in the normal way.

After breakfast collect your ring numbers, if you have not already done so, and where necessary take your horse or pony to be measured. After this give your animal a final brush over. For animals which are normally kept out at grass, a very satisfactory result can be obtained without removing any grease from the coats by running a slightly damp body brush over the surface of the coat, always in the direction of the hair. This gets rid of all the surface dust from the coat. As soon as the hair is dry, the coat is polished up to its maximum shine with a coarse stable rubber.

Any white leg markings can be made shining white by rubbing them over with a block of white dog chalk – obtainable from most pet stores. This removes any trace of yellow from the hair. The hooves should then be oiled.

When the horse is quite ready the person who is going to lead him in the show ring should get changed. The correct ring number must be tied round the leader's waist; where more than one horse is being shown, all the numbers can be tied on, one on top of the other in class order, which will save time later. At some shows it is customary to tie the ring numbers on to the browband of the bridle, in which case specially designed numbers are supplied. In the case of carriage and heavy horses it is customary to tie the numbers round the individual horses' necks.

The bridle is put on the horse and the tail bandage removed. Wait in your loose-box until the first call is made over the loud speakers or by the horse foreman for the entries in your class to go down to the collecting ring. Make your way down slowly; there will be plenty of time. The collecting ring steward will tell you when to go into the ring for judging. If you have never shown a horse in-hand before, it would be better if you allowed one or two seasoned exhibitors to go in front of you. Otherwise it is quite a good idea to lead the class into the ring yourself then, especially if your animal is a good walker, you can set the pace. Otherwise if possible place your horse between two smaller animals or ones which are obviously inferior so that the comparison is in your favour.

Walk out smartly on the right rein. If the horse in front of you is walking slowly go on a slightly wider circle, but do not let the horse get obscured from the judge's view.

The judge will now be assessing his class and he will probably soon start to pull the horses he likes best into the middle of the ring – keep half an eye on the judge and be ready to come into the centre of the ring should the ring steward call you. When you are pulled in stand your horse up facing the grandstand. When the whole class is in line, the judge may examine each horse, or he may only do this when the horses are pulled out of line. However, whenever you see the judge approaching, make absolutely certain that your horse is standing with his weight evenly distributed on all four legs. Wait until the ring steward tells you to bring your horse out for the judge to see, then stand him sideways on to the crowd and in front of the judge. Make sure he is standing evenly and that his hind leg furthest away from the judge is slightly forward.

When the judge has finished looking at your horse he will tell you to walk to the far end of the ring, turn and trot back past him to the other end of the ring. As you go up the ring make sure your horse is walking out to the best of his ability. As you turn, walk round your horse, never ask him to walk round you otherwise he will tend to become unbalanced. For the same reason, always walk two or three strides after you have turned before giving the command to trot. Be ready to give him a tap with your stick if he fails to get the message.

Providing you have a horse which moves really well, you can ask him to trot out fast. The biggest danger here is that the horse will lean on the bit, go on his forehand and sprawl behind. Therefore once your horse is trotting well (some horses seem to 'change gear' as they lengthen their stride) drop your hand so that the horse is going on a slack rein. In this way he will have to carry himself and not lean on the bit. Some Arabian and pony judges require you to trot once round the ring, but hunter judges only require you to go the length of the ring. Just before pulling up give the word of command to slow down as, if obeyed, this will avoid you having to pull on the horse's mouth.

After your show, return to your place in the line and be ready to stand your horse up again should the judge come back down the line for a final look. Most judges, at this stage, ask the whole class to walk round the ring again in a somewhat reduced circle. They then call the exhibitors into their final order of judging and the rosettes are handed out. Even if you are disappointed it is only polite to thank the judge as he hands you the rosette and card.

When you have been judged make certain that you are not required again for any championship or special prize before returning to the horse lines. All first and second prize-winners in youngstock classes are expected to compete for the breed championship; if there is more than one foal class, then they too compete for their own championship.

At most of the larger shows a grand parade of prize-winners is staged. Failure to parade usually means the withdrawal of any prize money won. A steward is posted on the collecting ring or main ring gate to take the numbers of all exhibitors who do parade. This grand parade is usually held in the late afternoon of the day of judging, or at some shows, on the next day, which means if you have won a monetary prize you must stay at the show for another night, in which case there is no extra charge for stabling.

It is customary not to plait up for a parade and indeed, most people remove their plaits before the parade so that they can load up and go home immediately afterwards. Non-prize-winners are usually at liberty to go home straight after the judging although some shows keep their gates shut until about 4 p.m. so that the general public can see all the horses.

If you have won a major prize a steward may come to enquire about your horse's past show record. This information is for the commentator during the grand parade and will be read out as your horse enters the ring.

When you are getting ready for the parade fix your horse's rosettes as far as possible to the near side of the bridle, so that the crowd can see them as you walk round. Any ribbons which might flap into the horse's eye should be anchored underneath the bridle so that they don't frighten him. All rosettes from every class won at that particular show should be worn for the parade; otherwise during the judging only rosettes won in the particular breed or type classes should be worn, regardless of whether you are still showing under the same judge.

About half an hour before the time to parade you will be called down to the collecting ring, where you will be sorted out into class order: the breed or type champion and reserve champion, from whichever class they came, usually lead the parade of that breed or type, followed by the prize-winners in order, starting with the first prize-winner of each class, followed by the second and third, etc.

All the prize-winners usually walk once round the main ring and then line up across the middle of the ring. The cups and trophies are then awarded, the winners going forward to receive them. After presentation of the trophies the prize-winners then file once more round the ring and out.

Small shows and gymkhanas seldom have veterinary surgeons on the gate to check the health of the horses coming on to the showground. However, they do have an official to tell you where to go and to collect entrance money from the general public.

Your horse-box should be taken to the parking space allotted to these vehicles. This type of show seldom if ever provides stabling for its exhibitors, so the horses must remain inside the horse-boxes when not actually being shown or got ready for the ring.

Quiet horses can be tied to the side of the horse-box with a hay-net to keep them occupied, but this is not to be recommended for youngstock.

On arrival at the show, it is advisable to go to the secretary's tent to make your entry, if you have not already done so, and collect your ring numbers. Most larger shows state that foals must be at least one month old before they can be shown, but some smaller shows have no such regulations, and foals which are only a week old are sometimes seen in the ring. This practice is not to be recommended for two main reasons. The first is that at this age the foal is too young to be asked to walk several times round a show ring while it is being judged. It will get tired and this will lower its resistance to disease. Secondly, at a show it will be exposed to the risk of various diseases, to some of which it will probably not have any immunity. Therefore *foals which are less than one month old should not be shown.*

For one-day shows it is absolutely essential that you should arrive at least one hour before you are due in the ring. If your horse will stand quietly while you plait it up at the show, it is a good idea to leave the tail until you arrive on the showground, as tails can easily be rubbed out while the horse is travelling. The mane is probably better put in the night before the show or early in the morning before you leave home. There are two disadvantages to plaiting manes the night before the show: horses tend to rub their manes and necks when they are left plaited up over night, so that, even if the plaits don't actually come undone, they will look rather wispy by the morning; also small pieces of straw tend to get entangled with the plaits when the horse lies down and these have to be picked out carefully in the morning. However, if plaiting up is left until arrival at the showground, it may mean a very rushed job, as there can be traffic hold ups in the immediate vicinity of the showground.

At a small show it is unwise to rely on an official to tell you when to come into the ring for judging, so a very close watch must be kept on the progress of the ring events, otherwise you might miss your class. Many small shows do not have a collecting ring and the entries file straight into the judging ring.

The turn-out and method of judging and showing at small shows is identical to that already described for larger shows, except that small shows don't usually organise a grand parade later in the day, so exhibitors are free to go home immediately after the judging of their classes if they so wish.

16

INSURANCE OF YOUNGSTOCK

Foals up to 24 hours old represent the greatest insurance risk. As they get older, so the risk lessens, but youngstock, particularly from the time they are weaned, are more liable to accidents and injuries than older horses. Therefore it usually pays to insure all young horses.

Basically, a large proportion of horse insurance is placed through underwriters into Lloyds. Not all underwriters will deal with this form of insurance, but there are some who do, and a very few who make it their speciality.

Non-Thoroughbred horses are usually accepted at a higher rate than Thoroughbreds, possibly due to the normally higher standard of management in most Thoroughbred establishments.

FOAL INSURANCE

It is possible to insure an unborn foal subject to a normal positive preganancy test and including the risks of twinning. This insurance is offered with the view to recovering the stallion fees, etc., in the event of a mare certified in foal not producing a live foal which lives for 30 hours.

The mare's previous breeding record influences the rate charged. Maiden mares and those with clean records command the lowest rate; mares which have slipped foal once are charged a proportionately higher rate; other mares may not be accepted for unborn foal insurance.

Foals are not accepted for insurance until they are 24 hours old; if not insured until seven days old there is an approximate 0.5 per cent reduction in the rate. Foals over two weeks old, yearlings and older horses are normally assessed at about 1.0 per cent less than the newborn foal rate.

HEALTH REQUIREMENTS

It is a condition of insurance that the animal is fit and well at the inception of cover. Therefore, unless an animal has been bought at public auction, when the insurers are happy that the said animal will be in good health, they require a declaration of health form signed by the owner or his agent, or a veterinary certificate of health. Animals with an insurance value of over £3,000, or £10,000 on renewal, must have a veterinary certificate, and until receipt of this by the insurance company are only covered for death by accidental injury.

VALUATION

Some people have difficulty in placing a value on a newborn foal. The main problem is to work out a figure which will cover the stud fee involved and the keep of the foal for a year. The keep of the mare can be included but might be better allocated to the mare herself for insurance purposes. An overall figure which is in the region of three times the original nomination fee is usually acceptable, but in most cases a lower figure is more realistic, particularly where a high nomination fee is involved. Over-insurance is no advantage. As a general rule of thumb, foals can be estimated as being worth double the value of the nomination and yearlings double the foal valuation.

RATE REDUCTIONS

Considerable reductions can be effected in insurance rates if an owner has a number of horses, as a stud owner does. The reduction will vary according to the total value of the horses insured.

Other reductions can be obtained by joining certain clubs or groups which offer insurance at reduced rates as part of the perks of membership of the club or group. This can amount to a 15 per cent reduction in rates.

POSSIBLE EXTENSION TO EXISTING POLICIES

Veterinary fees: for fees incurred as a result of an accident, sickness or disease (not including foaling or inoculation), up to £50 any one claim but excluding the first £5 of each and every claim.

Theft (including straying) an additional rate of approximately 0.5 per cent is usually charged.

Loss of use: upon payment of an extra premium, the insurers will pay up to 60 per cent of the sum insured in the event of the insured animal suffering a permanent disability as a direct result of illness or disease and/or injuries sustained accidentally resulting in its being unable to fulfil the functions for which it is kept, the disability being such that destruction is not required for humane reasons.

Economic slaughter: as an alternative to 'loss of use', an economic slaughter extension can be obtained, which pays up to 75 per cent of the sum insured if the animal is destroyed for economic reasons, with the option of the owner electing to be paid 50 per cent of the sum insured should he wish to keep the animal, providing underwriters liability ceases immediately.

Short-term insurance: cover can usually be obtained for short-term periods, which are often times of maximum risk. Most eventualities can be covered including foaling risks, all surgical operations, including castration, and all forms of transit whether by sea, road or air, although *some* of these risks are usually already covered by a 12-months policy. Short-term policies can be taken out for uninsured animals.

Claims: these are generally straightforward providing one is dealing with a reputable company and as long as all the requirements have been met at the start of cover. Under these circumstances it usually only takes about two weeks from the receipt of a *post-mortem* certificate to settle a claim.

The insurance company should be notified in the event of a likely claim arising from a serious accident or illness. Should an animal die or meet with a fatal accident the insurance company must be notified immediately and the carcase not disposed of until it has been cleared by the insurance company. They will normally require a *post-mortem* report.

APPENDIX A

USEFUL INFORMATION AND ADDRESSES

Qualifications in stud work

For anyone wishing to gain qualifications in stud work the following schemes are available in the United Kingdom:

(1) The Newmarket School of Stud Management, Derisley Wood Stud, Newmarket, Suffolk, offers a course which includes a stud season and combines practical work with lectures by some of the leading authorities in the Thoroughbred industry. A certificate is awarded to successful candidates.

(2) The National Pony Society, 85 Cliddesden Road, Basingstoke, Hants., organises examinations for candidates with one year's or three years' continuous stud experience – they also keep lists of studs which are normally prepared to accept working pupils. The examinations are for the Diploma in Pony Mastership and Breeding and for that of Stud Assistant. Both can be taken with or without riding. Details of the syllabus can be obtained from the National Pony Society at the above address.

(3) The English and Irish National Studs accept working pupils for a stud season. Those interested should write to: The Director, The English National Stud, Newmarket, Suffolk; or The Manager, The Irish National Stud, Tully, Co. Kildare, Republic of Ireland.

Useful addresses

Messrs. Weatherbys (*General Stud Book*),
Sanders Road,
Wellingborough,
Northants NN8 4BX.

Hunters Improvement & National Light Horse Breeding Society,
8 Market Square,
Westerham,
Kent.

Tindall & Son (specialists in stud stationery and printing),
50 High Street,
Newmarket,
Suffolk.

R. P. Faulkes Esq. (specialist horse photographer),
88 High Street,
Newmarket,
Suffolk.

Stud and Stable,
59 High Street,
Ascot,
Berks.

Bloodstock and General Insurance Brokers,
20 High Street,
Newmarket,
Suffolk.

The National Foaling Bank,
Meretown Stud,
Newport,
Shrophsire. (Newport 811234)

Horse Requisites (Newmarket) Ltd.,
The Railway Station,
Newmarket,
Suffolk.

APPENDIX B

SYMPTOMS GUIDE TO CHAPTER 3

Below is a quick guide to the symptoms of the various conditions which can affect young foals.

Symptom	*Condition or disease*	*Reference* (Page)
Above-normal temperature	Barkers and dummy foals	37
	Scouring	38
	Influenza	44
	Joint ill/navel ill	44
	Meningitis	46
	Pneumonia	47
	Salmonella	50
	Septicaemia	50
	Strangles	52
	Swamp fever	40
Rapid breathing/respiratory distress	Barkers and dummy foals	37
	Haemolytic foals	41
	Influenza	44
	Joint ill/navel ill	44
	Meningitis	46
	Pneumonia	47
	Ruptured bladder	50
	Salmonella	50
	Septicaemia	50
	Strangles	52
Foal off suck	Barkers and dummy foals	37
	Haemolytic foals	41
	Joint ill/navel ill	44
	Meconium retention	45
	Meningitis	46
	Pneumonia	47
	Ruptured bladder	50
	Salmonella	50
	Septicaemia	50
	Sleepy foal disease	51
	Strangles	52

Lameness	Bruised sole	—
	Joint ill	44
	Round joints	48
	Tetanus	52
	Wobblers	55
Inco-ordination of limbs	Wobblers	55
Swollen joints	Joint ill	44
	Round joints	48
Swellings in the angle of the jaw	Guttural pouch tympanitis	41
	Herpesvirus respiratory infection	48
	Snotty noses	51
	Strangles	52
Nasal discharge	Herpesvirus respiratory infection	48
	Influenza	44
	Pneumonia	47
	Snotty noses	51
Cough	Strangles	52
	Influenza	44
	Snotty noses	51

INDEX